T0211644

A Deeper Perspective on the Fundamentals of Digital Communication, Security, and Privacy Protocols

This book, divided into three parts, describes the detailed concepts of Digital Communication, Security, and Privacy protocols.

In Part One, the first chapter provides a deeper perspective on communications, while Chapters 2 and 3 focus on analog and digital communication networks.

Part Two then delves into various Digital Communication protocols. Beginning first in Chapter 4 with the major Telephony protocols, Chapter 5 then focuses on important Data Communication protocols, leading onto the discussion of Wireless and Cellular Communication protocols in Chapter 6 and Fiber Optic Data Transmission protocols in Chapter 7.

Part Three covers Digital Security and Privacy protocols, including Network Security protocols (Chapter 8), Wireless Security protocols (Chapter 9), and Server Level Security systems (Chapter 10), while the final chapter covers various aspects of privacy related to communication protocols and associated issues.

This book will offer great benefits to graduate and undergraduate students, researchers, and practitioners. It could be used as a textbook as well as reference material for these topics. All the authors are well-qualified in this domain. The authors have an approved textbook that is used in some US, Saudi, and Bangladeshi universities since Fall 2020 semester – although used in online lectures/classes due to COVID-19 pandemic.

A Deeper Perspective on the Fundamentals of Digital Communication, Security, and Privacy Protocols

Kutub Thakur, Abu Kamruzzaman, and
Al-Sakib Khan Pathan

CRC Press
Taylor & Francis Group
Boca Raton   London

CRC Press is an imprint of the
Taylor & Francis Group, an **informa** business

First edition published 2022
by CRC Press
6000 Broken Sound Parkway NW, Suite 300, Boca Raton, FL 33487–2742

and by CRC Press
4 Park Square, Milton Park, Abingdon, Oxon, OX14 4RN

CRC Press is an imprint of Taylor & Francis Group, LLC

© 2022 Taylor & Francis Group, LLC

ISBN: 978-1-032-29287-8 (hbk)
ISBN: 978-1-032-29292-2 (pbk)
ISBN: 978-1-003-30090-8 (ebk)

DOI: 10.1201/9781003300908

Typeset in Times
by Apex CoVantage, LLC

To my father Billal Uddin Thakur, my mother Nurun Nehar Parul and my wife Nawshin Thakur

– Kutub Thakur

To my respected father Abu Ahmed, my beloved mother Jahanara Begum and my devoted wife Nicole Woods

– Abu Kamruzzaman

To my parents: my father, Abdus Salam Khan Pathan and my mother, Delowara Khanom

– Al-Sakib Khan Pathan

Contents

Preface

The intent of writing this book is to offer an "easy-read" volume that makes complicated issues about communication, security, and privacy protocols easily accessible even to the general readers. While it can be used as a textbook, various parts can also be used for research works in academia or industry. This book is divided into three parts describing the detailed concepts of Digital Communication, Security, and Privacy protocols.

Part One is entitled *Introduction to Analog and Digital Communication Protocols*. In this part, the first chapter provides a deeper perspective on communication while Chapters 2 and 3 focus on analog and digital communication networks.

Part Two is entitled *Details of Digital Communication Protocols*. In this part, the first chapter (Chapter 4) highlights the major Telephony Protocols, Chapter 5 focuses on important Data Communication Protocols, Chapter 6 discusses Wireless and Cellular Communication Protocols, and as the last chapter, Chapter 7 presents Fiber Optic Data Transmission Protocols.

Part Three of the book is on *Digital Security and Privacy Protocols* where the first chapter (Chapter 8) focuses on Network Security Protocols, then Chapter 9 explores Wireless Security Protocols, Chapter 10 covers the Server Level Security Systems, and the last chapter of this book (Chapter 11), discusses various aspects of privacy related to communication protocols and related issues.

We hope that the book will offer great benefits to graduate and undergraduate students, researchers, and practitioners. As mentioned, it could be used as a textbook as well as a reference material for these topics for research activities at various levels.

We are very grateful to Almighty Allah for giving us this time to be able to complete writing this book. We also thank our family members who always encourage us to excel in professional and research fields and provide continuous support.

Kutub Thakur, PhD
Abu Kamruzzaman, PhD
Al-Sakib Khan Pathan, PhD

Authors

Kutub Thakur, PhD, is Founding Director of NJCU Center for Cybersecurity and Assistant Professor and Director of Cybersecurity Program at New Jersey City University. He worked for various private and public entities such as United Nations, New York University, Lehman Brothers, Barclays Capital, ConEdison, City University of New York (CUNY), and Metropolitan Transport Authority. He earned his PhD degree in Computer Science with specialization in cybersecurity from Pace University, New York, MS in Engineering Electrical and Computer Control Systems from the University of Wisconsin, and BS and AAS in Computer Systems Technology from CUNY. He has worked as a reviewer for many prestigious journals and published several papers in reputable journals and conferences. His research interests include digital forensics, network security, machine learning, IoT security, privacy, and user behavior. Dr. Thakur is currently serving (also, served) as the program chair for many conferences and workshops. He is also currently supervising (also, supervised) many graduate and doctoral students for their theses, proposals, and dissertations in the field of cybersecurity.

Abu Kamruzzaman, PhD, has been an assistant professor (adjunct) teaching undergraduate- and graduate-level computing courses in various campuses in NYC since 2001. He is currently holding a full-time position in NYC under CUNY as an Enterprise Application Developer supporting SQL based Data warehouse and Microsoft.Net Web Enterprise applications infrastructure. Dr. Kamruzzaman has maintained multiple IT careers in multiple public entities in NYC since 2001. He has more than 20 years of professional experience building and leading projects for mobile applications, enterprise applications, web applications, databases, and data analytics. He is proficient in working with both Windows and UNIX environments enterprise systems. Dr. Kamruzzaman earned his PhD in Computer Science (CS) concentration Artificial Intelligence from Pace University, New York, USA; MA in CIS from Brooklyn College/CUNY, New York; BS in CS from Binghamton University/SUNY, New York; and AAS in CS from LaGuardia Community College. He built apps frequently used by 500K+ active users and 20M+ historic users. He also built dashboards and analysis as Senior Business Intelligence Architect in CUNYSmart team building the data warehouse for CUNY, which enrolls over 0.5M students per semester. Dr. Kamruzzaman has published papers in multiple journals including *Institute of Electrical and Electronics Engineers* (IEEE). He has presented and chaired in several conferences nationally and internationally. His research interests include data science, cyber security, machine learning, quantum computing, and cloud computing. He is the recipient of many awards for his extraordinary accomplishments. He also served in the dissertation committees for CS PhD students and a faculty of the DPS doctoral program at Pace University.

 **Al-Sakib Khan Pathan, PhD,** is Professor in the Computer Science and Engineering Department, United International University, Bangladesh. He is also serving as a PhD co-supervisor (external) at Computer Sciences Department, University Ferhat Abbas Setif 1, Algeria. He earned his PhD in Computer Engineering in 2009 from Kyung Hee University, South Korea, and BSc degree in Computer Science and Information Technology from Islamic University of Technology, Bangladesh, in 2003. In his academic career, he worked as a faculty member in the CSE Department of Independent University, Bangladesh, during 2020–2021, Southeast University, Bangladesh, during 2015–2020, Computer Science Department, International Islamic University Malaysia, Malaysia, during 2010–2015, at BRACU, Bangladesh, during 2009–2010, and at NSU, Bangladesh, during 2004–2005. He served as Guest Professor at the Department of Technical and Vocational Education, Islamic University of Technology, Bangladesh, in 2018. He also worked as a researcher at Networking Lab, Kyung Hee University, South Korea, from September 2005 to August 2009, where he completed his MS leading to PhD. Dr. Panthan's research interests include wireless sensor networks, network security, cloud computing, and e-services technologies. He is a recipient of several awards/best paper awards and has several notable publications in these areas. So far, he has delivered 32 keynotes and invited speeches at various international conferences and events. He was named on the List of Top 2% Scientists of the World, 2019 and 2020 by Stanford University, USA, in 2020 and 2021. He has served as General Chair, Organizing Committee Member, and Technical Program Committee member in numerous top-ranked international conferences/workshops such as INFOCOM, GLOBECOM, ICC, LCN, GreenCom, AINA, WCNC, HPCS, ICA3PP, IWCMC, VTC, HPCC, SGIoT, etc. He was awarded the IEEE Outstanding Leadership Award for his role in the IEEE GreenCom'13 conference and IEEE Outstanding Service Awards twice in recognition and appreciation of the service and outstanding contributions to the IEEE IRI'20 and IRI'21. He is currently serving as Editor-in-Chief of *International Journal of Computers and Applications*, Taylor & Francis, UK; Editor of *Ad Hoc and Sensor Wireless Networks*, Old City Publishing, *International Journal of Sensor Networks*, Inderscience Publishers, and *Malaysian Journal of Computer Science*; Associate Editor of *Connection Science*, Taylor & Francis, UK, *International Journal of Computational Science and Engineering*, Inderscience; Area Editor of *International Journal of Communication Networks and Information Security*; guest editor of many special issues of top-ranked journals; and editor/author of 28 books. One of his books has been included twice in Intel Corporation's Recommended Reading List for Developers, second half 2013 and first half of 2014; three books were included in IEEE Communications Society's (IEEE ComSoc's) Best Readings in Communications and Information Systems Security, 2013, several other books were indexed with all the titles (chapters) in Elsevier's acclaimed abstract and citation database, Scopus and in Web of Science (WoS), Book Citation Index, Clarivate Analytics, at least one has been approved as a textbook at NJCU, USA, in 2020, one is among the

Top Used resources on SpringerLink in 2020 for UN's Sustainable Development Goal 7 (SDG7) – Affordable and Clean Energy, and one book has been translated to simplified Chinese language from the English version. Also, two of his journal papers and one conference paper were included under different categories in IEEE ComSoc Best Readings Topics on Communications and Information Systems Security, 2013. Dr. Panthan also serves as a referee of many prestigious journals. He received awards for his reviewing activities including one of the most active reviewers of IAJIT several times and Elsevier Outstanding Reviewer for Computer Networks, Ad Hoc Networks, FGCS, and JNCA in multiple years. He is Senior Member of the IEEE, USA.

PART ONE

Introduction to Analog and Digital Communication Protocols

Evolution of Communication Protocols

1

WHAT IS A COMMUNICATION PROTOCOL?

A communication protocol is a set of predefined rules and laws used for establishing meaningful communication between sender and receiver of the signals (via the transmission of signals or information from one communication node to another communication node). The communicating nodes are commonly referred to as the transmitter and the

DOI: 10.1201/9781003300908-2

receiver.[1] Telecommunication protocols play a vital role in modern communication technologies, such as:

- Telephone systems
- Television systems
- Satellite systems
- Data communication
- The Internet technologies
- Brain–computer interface
- Tele-learning systems
- Radar systems
- Cellular phone systems

All the aforementioned communication systems interact with the sender and the receiver of the signals with the help of numerous communication protocols. The natures of those communication protocols are different from each other, but the core principles and objectives of all those communication protocols are the same. If you look into the communication system with a deeper perspective, you will find that there are many communication protocols involved in accomplishing modern communication through multiple nodes of communication networks spread across the continents. At every point of time, the conversion of protocols and the change of the state of signals occur, which are handled by different protocols associated with signal processing and communication.

Physically speaking, a communication system consists of a sender, a receiver, and a communication channel on which the signal or message is transmitted from one sender to a receiver of the signals or message. But in real-world communication systems, the number of components of a communication system is six as listed in the following:[2]

- **Terminal nodes** – The terminal nodes also referred to as *terminals* are the receiving and sending elements in the communication system. In other words, we can say that the sender device and the receiver device are the terminals of the communication systems. The terminals can also be named input devices and output devices in modern telecommunication systems.
- **Communication channel** – For transmitting electrical or electromagnetic signals from one point to another one, a channel of communication is required. The channel of communication is built on certain wires, vacuums, and air medium in modern communication systems. Examples of communication media include twisted pair, coaxial cables, fiber, microwave, extra-terrestrial ultrahigh frequency (UHF), optics, light, and so on.
- **Message** – Message is the information coded into the signals, which is transmitted from one terminal of communication to the other one. The message may consist of image, video, audio, or text.
- **Communication protocols** – Communication protocols are the most fundamental component of establishing a meaningful communication between two terminals of the communication system. It is a set of predefined rules and responses of the signals sent from one node to the other one. The entire process of transmitting of message from sender to receiver is done by communication protocols.

- **Communication processors** – A communication processor is another very important part of the communication system, which plays a very critical role in processing the control, feedback, and other additional supportive functions required for establishing a smooth communication across different types of network elements other than transmitters and receivers. Examples of communication processors include signal converters, boosters, signal inverters, multiplexers, dividers, and the like.
- **Control software** – A software platform that is used for managing, configuring, controlling, and performing other operations and maintenance activities on the communication systems is known as control software. This software also communicates between two elements of nodes through certain communication protocols.

As discussed earlier, a communication protocol is like spine in the entire communication system; so, any kind of snag or glitch in the communication protocol will lead to the collapse of meaningful communication. There are so many examples of communication, and a few examples of modern communication protocols are listed as follows.

- Transmission Control Protocol/Internet Protocol (TCP/IP)
- Session Initiation Protocol (SIP)
- Signaling System No. 7 (SS7) Protocol
- Time-Division Multiplexing (TDM) Protocol
- And many more

HISTORY OF COMMUNICATION PROTOCOLS

The history of telecommunication protocols is as old as sending remote signals in interlocation communication. The start of the communication protocols dates back hundreds of years when people started using smoke and fire signals for remote communication. Later, the drums were used for remote communication through different types of drumbeat combinations. The use of voice and visual signals were the only two major sources of remote communication at that time.

Slowly and gradually, the proper systems for remote communication were established. Among such preliminary telecommunication systems, semaphore is very important to name. The first semaphore system emerged in Europe in the 1790s. The antenna pattern of semaphore is an example of communication protocol.[3] Electricity was invented at the beginning of the 1800s and evolved through a couple of decades until people started using it as communication signals. In the 1790s, visual telegraphy based on clock hands of a pair of clocks was introduced. The formation of the hands of the clocks in different formats was an example of communication protocol in that visual form of telegraphy.

Later, Samuel Thomas von Sömmerring created the first electrochemical-based telegraph in 1809. In this telegraphy, the current would electrolyte the chemical solutions and a series of air bubbles would be produced at the receiver. The series of bubbles was used as the codes of communication. And, this entire system of bubble signals and their understanding is an example of communication protocol.

Samuel Morse developed a communication protocol for a new and advanced version of electrical signal-based telegraphy in 1837. In this advanced telegraphy, the Morse code was used as the communication protocol. Thus, the advancements continued in the communication protocols. The patent for the electric telephone was awarded to Alexander Graham Bell in the US in 1876. This phone would use a liquid transmitter for sending voice information from one telephone set to another one. The oscillating current/voltage was used for transmitting signals. The name of oscillating signals is now referred to as amplitude modulation (AM) of the input signal over current. Later, the same AM was used for transmitting the voice signals over radio waves.

The other examples of different communication protocols used for telephone communication include TDM, SS7, Channel-Associated Signaling (CAS), SIP, and so on. The advancement of communication protocols continued not only in voice communication over telephone, television, and radar systems but also in data transmission. Many of the data transmission protocols started after the middle of the 20th century.

Frequency modulation (FM), pulse modulation, and other protocols were used in television and satellite communication. The microwave and UHF were used in the other extra-terrestrial communication. Numerous communication control protocols were deployed in the wireless communication arena. The revolution in interactive communication started with the advent of data communication networks pioneered by the Advanced Research Projects Agency Network (ARPANET) and TCP/IP in 1983.[4] The creation of the world wide web (WWW) by Tim Berners-Lee in 1990 became the core protocol for the Internet, which revolutionized the communication systems in the world.

Many cellular mobile and wireless technologies powered by numerous modern communication protocols have changed the landscape of modern telecommunication. Among such communication protocols and technologies, 3G, 4G, 5G, Bluetooth, ZigBee, IoT, Multiple-Input and Multiple-Output (MIMO), and numerous types of wireless modulation mechanisms are some of the very important ones.

TYPES OF SIGNALS USED IN COMMUNICATION HISTORY

Communication is the name of transmission of a meaningful signal from one person to another one. Similarly, from the technological aspects, communication is the transmission of a meaningful signal from one point and the reception of the same meaningful signal at the other or receiving end. For the purpose of communication, many types of signals were used in the long history of telecommunication. A few very important types of signals used in history for different types of communications powered by different kinds of protocols are mentioned in the following.

Visual Signals

The visual signal was one of the most popular signals for short-range communication. The waving of some objects like arms, piece of cloth, smoke, or any other similar type of stuff would help transmit signals from one location to other where the normal voice would not reach easily. The technical system that used visual signals for telecommunication was the semaphore created by Claude Chappe in 1792.[5] The semaphore is also known as visual telegraphy. The Chappe telegraphy used a special pattern of objects to send information from one tower to another (or, to the other location).

Voice Signals

The sound signal is the most common type of signal extensively used for communication from its beginning. The sound signals included the voice created by the vocal cords. The very basic protocol of voice communication was based on the creation of different types of emotional sounds like crying, singing, screaming, yelling, laughing, talking, murmuring, and other forms. But the outreach of these types of voice signals was very limited up to a few meters or a few hundred meters. To use the voice in a more effective way, people created drums and lung-blown manual horns to carry the voice longer distances.

Different types of drums were traditionally used in numerous African countries, Latin American countries, China, and other parts of the world for communication purposes.[6] Different patterns of drumbeats were predefined for certain messages to convey to the remote locations. The use of divine conch shell or sacred conch shell (Shankha) was another very common tool for voice-based telecommunication of signals for a longer distance in ancient India. The common type of conch used in creating voice signals is shown in Figure 1.1.

FIGURE 1.1 Conch for creating sound (Pixabay).

Smoke Signals

Smoke was another very important form of remote signaling in ancient telecommunication systems. Smoke was extensively used by ships, boats, and stranded people for getting help. But certain patterns for smoke signals were also developed for sending meaningful signals to remote locations. This is a form of visual communication, which is only effective in the daytime when smoke can be seen. In the ancient periods, Chinese soldiers would send a smoke signal through a series of towers located along the Great Wall to send a message about any kind of threat or any preemptive measure for the army located along the Great Wall. They would send the message at about 750 km distance within a few hours.[7]

A complex system of signals based on smoke was created by Polybius in 150 BCE, and the Chinese Zhou Dynasty used smoke signaling extensively in their respective telecommunication systems at that time. Similarly, the native people in North America also used different types of smoke patterns for their communication systems for their respective tribes. Smoke is also used in the present-day (modern era) in certain limited applications of communication. The church of Vatican City still uses the smoke for their traditional communication messaging.

Fire Signals

Fire was one of the most important types of communication signals in the nighttime during the ancient periods. A very creative system of beacon lights was developed by the Greeks in the 3rd century BC. In this fire signal system, a beacon consisting of two walls built at a prominent position on a mountain would be used to send messages to another mountain about 100 km away. Every wall consisted of two portions that would be built in parallel columns. Five people with beacons would form a system of binary-like combinations of signals. Different forms of signals would be sent according to the given patterns to send a specific message to the other site. This system was designed and developed by Kleoxenos and Dimokleitos.[8] The pattern used in that fire communication was the example of communication protocol, which would provide the way for meaningful telecommunication between the two designated sites.

Electrical Signals

Electrical signals mark the start of the first revolution in telecommunication systems as well as in numerous other businesses, processes, and industries globally. Electricity was invented in 1800 when Volta invented the battery.[9] Later on, the electrical signals were not used for communication purposes until Samuel Morse developed an electric code that would be used in the electrical telegraphy for communication. Morse code is the first type of model communication protocol, which was developed in the 1830s.[10] It was a comprehensive coding system for electrical telegraphy. The evolution of the electrical telegraphy system continued across North America, Europe, and other parts of the world for about 100 years.

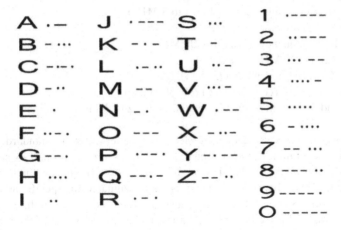

FIGURE 1.2 Morse code (Public Domain Pictures).

Morse code was coded with the dashes and dots. The basic Morse code is shown in Figure 1.2. For any emergency or help, Save Our Souls (SOS) was used in the Morse code. The SOS is simple to write: three dots for S and three consecutive dashes for O. The SOS word is still used in our informal communication even in today's so-called modern world (though *modernism* is a continuous process as in every second, we add to our past world!).

Electromagnetic Waves

Electromagnetic waves are a wide range of electromagnetic frequencies that can travel in the air as well as in a vacuum. Electromagnetic waves are created by a variation of electric fields, which creates the varying magnetic field. The creation of such waves associated with the electric and magnetic field is known as an electromagnetic field. The electromagnetic waves are classified into two domains, which are also interlinked with each other. Those two domains are wavelength and frequency. Both frequency and wavelengths are also used to classify electromagnetic waves.

This is very important to note that the frequency and wavelength are reversely proportional in terms of value. The combined range of frequencies or wavelengths of electromagnetic waves is also referred to as an electromagnetic spectrum. The electromagnetic spectrum is divided into different frequency bands. The electromagnetic waves are classified into the following categories of frequencies or wavelengths.[11]

- Tremendously low-frequency band (below 3 Hz)
- Extremely low-frequency band (3–30 Hz)
- Super low frequency band (30–300 Hz)
- Ultralow frequency band (300–3,000 Hz)
- Very low frequency band (3–30 kHz)
- Low-frequency band (30–300 kHz)

- Medium-frequency band (300 kHz to 3 MHz)
- High-frequency band (3–30 MHz)
- Very high frequency band (30–300 MHz)
- UHF band (300 MHz to 3 GHz)
- Super high frequency band (3–30 GHz)
- Extremely high frequency band (30–300 GHz)
- Tremendously high frequency band (300 GHz to 3 THz)

The above-listed classification of electromagnetic waves is standardized by the International Telecommunication Union (ITU). This is very important to note that any frequency, which is above 1 GHz, is known as microwave frequency due to its micrometer wavelength. Every band of frequency is also designated for specific applications in telecommunication. Some of those frequencies are reserved for research, while the other bands are designated for medical use and a few bands are designated for extra-terrestrial communication.

Modern Light Signals

The modern light signals are generated by different types of diodes commonly called the LASER diodes. The light signals generated by the diodes are transmitted through fiber optic cables in telecommunication systems. The example of modern laser light is shown in Figure 1.3.

There are different types of fiber optic cables, which require different types of light signals with different strengths and configurations. Laser light is also used for lighting and entertainment purposes nowadays. As far as telecommunication systems are concerned, laser light is used for fiber optic communication in our modern telecom systems.

FIGURE 1.3 Laser light show (Unsplash).

OPEN SYSTEMS INTERCONNECTION (OSI) MODEL

Open Systems Interconnection Model, commonly referred to as OSI Model of communication, is a conceptual standard of data communication in which the entire communication steps and stages between the two end-user applications are distributed into different types of layers. Those layers of OSI models are not a particular protocol of communication but a reference layer for a particular type of communication to take place in the domain of that particular layer of OSI model.[12]

The main objective of developing the OSI model was to establish a standard concept of developing the communication protocols for different types of applications, which would be developed for data communication in the future. With the help of this model, the interoperability of different applications and software products could be achieved. Otherwise, the interoperability of numerous products in older times was a big problem for the growth of the technology. By developing this model, all vendors and developers of the software products and applications would be able to easily create interfaces at different layers of communication to interoperate the product with the other products or networks. A working software application that runs on our computers and mobiles is referred to as the seventh layer of the OSI model, which is also named the Application Layer. The details of the layers of the OSI models will be discussed in the following subtopics.

The OSI model was designed and developed by the International Standard Organization (ISO) in early 1983. After comprehensive discussion and enhancements, this model was approved and adopted as a standard conceptual framework for data communication protocols in 1984. The initial purpose of designing such a conceptual model was to focus on developing detailed specifications for different types of interfaces. But, it was much more difficult and would create speed breakers for developing new interfaces and applications for different functionalities. Hence, the entire team finalized to devise a conceptual model for the development of different types of interfaces and communication rules based on the core layered system of the OSI model.

The conceptual model of the OSI model is shown in Figure 1.4.

The OSI model of communication is formed of seven layers known as 1, 2, 3, 4, 5, 6, and 7. The first four layers ranging from 1 through 4 are known as lower layers, while the second part of layers ranging from 5 through 7 is referred to as upper layers. The function of the lower layers is to move data of the application and network around, while the upper layers are mostly concerned about the functional, presentational, and interactive features of the application data. The communication between these seven layers of OSI occurs on the principle of pass-on or handover of the data to the upper and lower layers of a particular layer. Each layer of the OSI model performs a certain function to process the data and passes the packet of data to the next layer. This way, the entire protocol of communication works.[13]

To understand the concept of the functioning of the OSI reference model, it is always a great idea to discuss the OSI layers in the reverse sequence starting from layer 7 and ending with OSI layer 1. So, let us start from OSI layer 7.

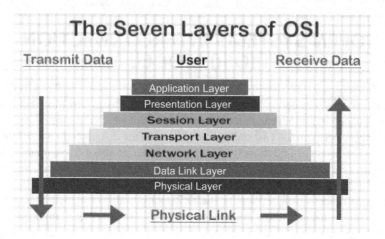

FIGURE 1.4   OSI model (Flickr).

Application Layer

Application layer is the topmost or the uppermost layer. It directly interfaces with the end-user, and at the lower end, it interacts with the session layer. Any web interface to show the application functions and take the input from the end-user is done in this layer of communication. It is also known as the seventh layer of the OSI model. The main source of interacting with the network resources by a software program or a manual user through the software interface is done in this layer. The user authentication, privacy, security of the application, quality of services (QoSs), and other such types of components of software application are identified and defined in this layer of the OSI model.

The main services provided by this layer of reference model through multiple protocols include the following:

- Sending and receiving emails
- Transferring file from one app to another one
- Remote accessing of any node or telnetting

The main protocols defined with reference to the application model include:

- Hypertext Transfer Protocol (HTTP)
- File Transfer Protocol (FTP)
- Telnet Protocol
- Simple Network Management Protocol (SNMP)

Presentation Layer

The presentation layer is the second topmost layer of the OSI reference model of communication. It is also referred to as OSI Layer 6. Technically speaking, the most important responsibilities of this layer of the OSI model are associated with the management of syntax and semantics. The meaning of those two terms is that this OSI layer takes care of the data received from the lower-level layer (session layer) in any format or syntax is changed in such a way that the application layer is able to understand and run. Similarly, the data sent from the application are changed in a format that is easily useable for the session layer (No. 6). In other words, we can say that it plays the role of the translator of the syntax and semantics used in the software protocol programming.

The major functions performed by the protocols in this reference layer of the OSI model include the following:[14]

- Translation of the formats required for all nodes, which are not compatible in terms of syntax and systems. It provides the data in the format that the desired machine understands.
- The plain text information of critical importance is encrypted and decrypted to safeguard the data during traveling and processing.
- The compression and decompression of the data are done by the defined protocols in reference to the presentation layer of the OSI model.

The examples of the most important real-world standards and concepts in reference to the presentation layer of the OSI model are listed as follows:

- American Standard Code for Information Interchange (ASCII)
- Musical Instrument Digital Interface
- Joint Photographic Expert Group (JPEG)

Session Layer

The session layer is the third top layer of the OSI model of communication. It is also known as OSI Layer 5 of communication. The main functionalities of the communication protocols developed in reference to the session layer or Layer 6 of the OSI model include the following:

- Establishing communication between two applications
- Monitoring and controlling the entire session of the communication
- Termination of the communication session between two applications

The transportation of dialogues, conversations, controls, and management signals between the two applications is managed by this layer. The coordination of the entire communication is the core responsibility of this layer of the OSI model of communication.

A few examples of protocols that work in reference to the session layer of the OSI model include the following:

- Remote Procedure Call
- Network File System Protocol
- NetBIOS Protocol

Transport Layer

The transport layer is the first layer of the lower-layer category of the OSI model of communication. It is also referred to as Layer 4 of the OSI model. The main objective of the protocols pertaining to this layer of the OSI reference model is to establish an end-to-end connection between the source and destination machines. The major types of connections at the transport layer in modern data communication are known as connection-oriented link and connectionless link.

The major functions of the commonly used protocols in reference to the transport layer of the OSI model include:

- Defining the address of the service points commonly referred to as the network port in the modern communication
- Fragmenting and defragmenting the data packet received from the lower-level protocols dealing with the network layer of the OSI model
- The error of the packet is controlled by the connection-oriented form of the protocol used in the transport layer of the OSI model
- The control of the flow of the data from source to destination computers
- Controlling connection in terms of establishing a reliable connection, which is free of packet loss and also a connectionless transmission of a packet with a high level of performance

The examples of the major protocols used in reference to the transport layer of the OSI model are listed as follows:

- TCP
- User Datagram Protocol (UDP)
- Sequenced Packet Exchange for Novell NetWare

Network Layer

Network layer is also referred to as Layer 3 of the OSI model of communication. It is the core layer in modern data communication, especially the Internet. All protocols that operate within the jurisdiction of the network layer in the OSI reference model of communication perform switching and routing functions to the destinations. The entire packet-switched network is fully dependent on the protocols dealing with the network layer. The network protocols take the message from the lower-level protocols referencing to data link layer and divide it into small packets and add source and destination addresses with the help of other auxiliary protocols and send the packet to the destination. In the reverse flow, the data packet received from the network is sent to the hardware addresses of the elements in the network under its jurisdiction.

The main functions of the protocols working under the jurisdiction of the network layer of the OSI model of communication include the following:[15]

- Converting logical address into physical address and vice versa
- Handling routing and switching of the connected networks

- Handle error control and flow control at the Layer 3 communication
- Controlling the sequences of the packets reaching from multiple routes
- Combining the smaller packets received through multiple routes to form a message and the reverse of the process too
- Managing routing tables of the networks and manipulating the dynamic routing tables in terms of any change in the network system

A few very most important examples of the protocols that work under the jurisdiction of the network layer of the OSI reference model of data communication are mentioned in the following list:

- IP
- AppleTalk Datagram Delivery Protocol
- Novell NetWare Internetwork Packet Exchange (IPX)

Data Link Layer

Data link layer is also known as Layer 2 of the OSI model. This layer consists of the two sub-layers known as:

- Media access control (MAC) layer
- Logical link control (LLC) layer

The first part of the data link layer deals with getting permission to network resources and transmit the data frame generated by the LLC layer. The coordination and control of the frames and packets are done by the LLC layer. The summary of the functions of the entire data link layer is given in the following list.[16]

- Receives the data packets from the network layer and forms a data frame to be transmitted to the physical layer.
- The physical address is attached to the data frame.
- Synchronization of frames is done by the data link layer.
- Error detection and error control at the frame transmission are handled by the data link layer.
- Flow control of the data frames is done by this layer.
- The access control to transmit the data in a network environment is handled by this layer.

The examples of protocols that work within the jurisdiction of the data link layer are given in the following list:

- Point to Point Protocol (PPP)
- High-Level Data Link Control (HDLC) protocol
- Institute of Electrical and Electronic Engineering (IEEE) Ethernet 802.2 Protocol

- Fiber Distributed Data Interface (FDDI) Protocol
- Frame Relay Protocol
- Asynchronous Transfer Mode (ATM)

Physical Layer

The physical layer is also known as Layer 1 of the OSI model. It deals with the communication links, signals, and bitstream structures transmitted over different types of mediums (this form of plural of "medium" can also be used in this context, while the other form "media" is used in many other cases) of communication such as twisted pair cable, coaxial cables, air interfaces, fiber optics, and others. The signals used at the physical layer of OSI models include manual and digital electrical signals, wireless signals, and light signals.

The function of the physical layer at receiving computer network node is to take the bitstream of the signals from the sender of the message through wire, air, or fiber optic cable and remove the overheads added by the sender's physical layer and send the data packet to the upper layer known as the data link layer. And in the converse scenario, the physical layer takes the data packet from the upper OSI layer known as the data link layer and adds certain agreed overhead and then sends the stream of signals through physical links connected through cables, air, or network cards.[17]

The main functions of the protocols working in reference to the physical layer of the OSI model of communication include the following:

- Receiving frames from the data link layer and converting them into electrical bitstreams
- Adopting the proper presentation or the coding scheme of the bits so that both – sending and receiving machines – understand them easily
- Deciding the data rate of the bit transmission
- Defining the network topologies
- Adding the right configuration of line such as point-to-point or point-to-multipoint, etc.
- Synchronizing the transmission of the bitstreams
- Setting up the types of transmission modes such as duplex, simplex, half-duplex, and full-duplex between two nodes.

Physical layer network functions are performed by the network card, which is shown in Figure 1.5.

The examples of protocols designed in reference to the physical layer communication include the following:[18]

- RS232 Protocol
- Bluetooth Protocol
- Digital Subscriber Line (DSL)
- Universal Serial Bus (USB)
- Integrated Services Digital Network (ISDN)

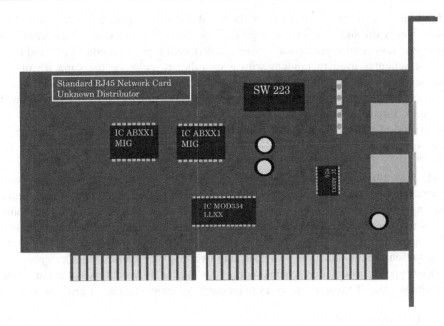

FIGURE 1.5   Network card (Pixabay).

The details of all those communication protocols, which work under the jurisdiction of different layers of the OSI model of communication, will be mentioned in the next chapters.

CATEGORIES OF MODULATIONS USED IN COMMUNICATION SYSTEMS

Modulation is one of the most crucial technologies that are the foundation of the entire telecommunication based on electric signals. At the beginning of radio transmission, the role of modulation was very important. In telecommunication technology, the modulation can be defined as:

> "The changing in any one or more than one component or characteristics of a harmonic waveform of an electromagnetic signal to get a meaningful output signal, which depicts the value of the desired transmitted signal".[19] The main reason of using the modulation technique in telecommunication is to transmit the original signal to a long distance through high-frequency carrier signals (which use high power levels).

All types of modulations used in modern telecommunication belong to the physical layer or Layer 1 of the OSI reference model of communication because all those modulations and demodulation protocols use the physical generation of manual or digital signals through oscillators, digital clocks, and other devices. All techniques used in wireless communication are based on different types of modulations and multiplexing schemes used in tandem with each other to enhance the capacity of the frequency band. For example, in the beginning when the radio signals were invented, they would use the unmodulated continuous wave (CW) signals for the transmission of the signals. Later on, Marconi and Fessenden used the modulated CW signals for the transmission of the signals. At that point in time, the generation of a high-frequency signal was also a bit problem and would need huge generators run by bulky electric power.[20]

The AM was also introduced by those scientists for carrying the original signal over the carrier band of very high frequency. AM is a type of analog signal modulation. In the beginning, there was no concept of any digital type of signal. The digital form of modulation was introduced with the advancements in electronic equipment and transmission technologies.

At present, the modulation is divided into two major categories. Each category has numerous types of modulation schemes commonly used in different types of communication technologies. The major categories of modulation/demodulation are given as follows:

- Analog modulation
- Digital modulation

The details of these categories of telecommunication modulation schemes are mentioned in the following sub-sections.

Analog Modulation

Analog modulation is a technique or process in which low-frequency radio signals such as voice and video, commonly referred to as baseband signals, are transmitted by superimposing them on high-frequency carrier signals to transmit them for a longer distance.[21] The main cause of using modulation is to transmit signals at longer distances and use a larger band of frequency for effective use of energy transmission. The image of analog modulation (AM) is shown in Figure 1.6.

In Figure 1.6, the modulating (information) or the baseband signal that is supposed to be transmitted over the carrier signal is shown as a sine-wave form in the middle. The carrier frequency is modulated in terms of the amplitude of the information signal. Here, the carrier frequency is shown as the sine waves at the top, and the amplitude modulated signal is shown at the bottom of the figure. The modulated signal will be re-generated through the demodulation process at the receiving end, and the demodulated signal is then fed to the output devices for recovering the signal transmitted from the source of the signal in the telecommunication system.

Analog modulation is divided into many types of modulations, some of which will be discussed later.

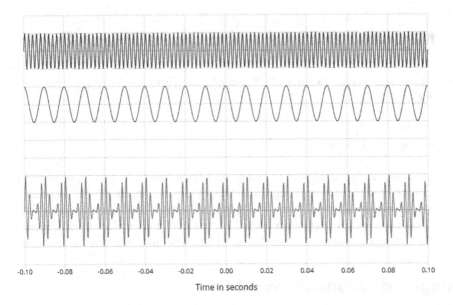

FIGURE 1.6 Basic amplitude analog modulation (generated with tool at: https://academo. org/demos/amplitude-modulation).

Digital Modulation

Digital modulation is the process of encoding the digital information signal, which is supposed to be transmitted to the remote receiver, in terms of phase, amplitude, and frequency of the carrier frequency (transmitted signal). Digital modulation is used for the better performance of carrier signals to achieve high-quality signal transmission with a minimum level of noise in the signal transmission.[22]

The major advantages of using digital modulation over analog modulation include the following:

- Higher level of transmission security
- Every single signal has multiple states to use
- Reduced noise in signals
- Easy to use different types of modern multiplexing techniques to increase the data transmission rate of the frequency band
- Easy to use multiple channels (multi-path) for better efficiency of the frequency band
- Support for numerous link communication conditioning schemes
- Increased power efficiency
- Lower bit error rate performance
- Reduced co-channel interference

MAJOR TYPES OF ANALOG MODULATIONS

Analog modulation is also known as CW modulation. There are two major types of analog modulations:

- AM
- Angle modulation

AM is the fundamental form of analog modulation, which was used primarily for the transmission of radio signals through radio transmitters and receivers. The angle modulation has two important components – frequency and phase – that can be used for modulation. Thus, two types of angle modulations are very common in telecommunication.

Types of continuous-wave or analog modulations are mentioned in the following.

Amplitude Modulation (AM)

This is the oldest form of analog modulation used in early-age radio communication. In this type of analog modulation, the message signal (baseband) such as voice or video signal is considered, which is a very low frequency signal and needs to be superimposed over the high-frequency carrier signal by modulating the amplitude or the strength of the carrier frequency in proportion to the amplitude or strength of the message signal.[23] The detailed explanation with an example of AM is shown in Figure 1.7.

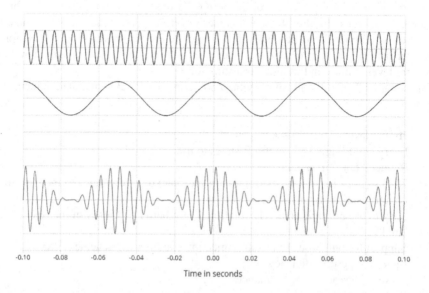

FIGURE 1.7 Amplitude modulation (generated with tool at: https://academo.org/demos/amplitude-modulation/).

The message signal of 20 Hz is shown in the middle of Figure 1.7 with a sine-wave form. The carrier frequency of 200 Hz is shown at the top as a sine wave. The modulated output signal through AM is shown at the bottom. You can see in the figure that when the amplitude of the signal of message/information (voice/video) increases, the strength of the carrier frequency also increases proportionally and when the amplitude of the message/information signal decreases, the amplitude of the carrier frequency also decreases. The frequency of the carrier signal is 10 times higher than that of the original message frequency. So, the strength to propagate the signal is 10 times greater than the normal strength of the signal.

Frequency Modulation (FM)

FM[24] is one of the two types of angular modulations, which is another very important part of analog modulation. Angular modulation has two types – one is FM and the other one is phase modulation (PM). The FM is a type of modulation in which the frequency of the carrier signal (high frequency) is changed in proportion to the amplitude of the message (or, modulating) signal, which is supposed to be transmitted through radio transmission (see Figure 1.8).

In this modulation, the frequency of the carrier signal increases as the amplitude of sine wave of the message signal increases. When the amplitude of the signal message starts decreasing after reaching the peak value, the frequency of the carrier signal also starts decreasing with the decreasing amplitude of the message signal. When the amplitude of the message signal becomes negative in the sine wave, the frequency of the carrier signal reduces further and reaches the minimum frequency when the message signal amplitude reaches the negative peak.

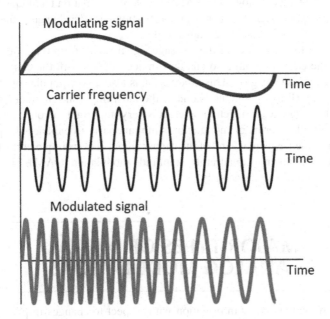

FIGURE 1.8 Frequency modulation.

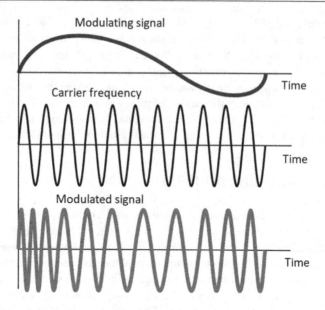

FIGURE 1.9   Phase modulation.

Phase Modulation (PM)

PM is a type of angle modulation in analog modulation. The PM technique is also used in digital modulation technology. The PM is a type of modulation technique in which the frequency of the carrier signal is changed in proportion with the change in the phase of the message (or modulating) signal (see Figure 1.9).[25]

The phase of the sine wave of the message signal varies from 0 to 360 degrees. The frequency of the carrier signal also changes in accordance with the changing value of the phase of the message signal. This modulation is also a very popular modulation. The phase shift keying (PSK) technique is one of the most important modulation techniques used in digital communication technologies. It is very important to note that PM is extensively used with other forms of modulation simultaneously to enhance the performance and efficiency of the frequency band of the carrier signal. The example of the combination of PM with AM is known as quadrature amplitude modulation (QAM). This modulation uses both PM and AM simultaneously.

MAJOR TYPES OF DIGITAL/ ANALOG MODULATIONS

Encoding the digital signal or information with respect to changes in phase, frequency, and amplitude to transmit the digital information is known as digital modulation. There are many popular types of digital modulations that are extensively used in modern telecommunication systems.[26] A few of those important types of digital modulation schemes are listed as follows:[27]

- Pulse amplitude modulation (PAM)
- Pulse code modulation (PCM)
- Pulsewidth modulation (PWM)
- Amplitude shift keying (ASK)
- Frequency shift keying (FSK)
- PSK
- QAM

Pulse Amplitude Modulation (PAM)

PAM technique is used in both digital and analog modulations. In a digital PAM modulation, the sampling approximation is done; otherwise, the basic technique is the same. In the PAM technique, the input modulating signal or message signal is modulated with a continuous train of pulses for a very short fraction of time. The schematic diagram of the modulating and modulated signals is shown in Figure 1.10.

The output signal or modulated signal is the multiplication of the amplitude of the message signal and pulse signal at a certain fraction of time when the pulse signal is turned on.[28] If the amplitude of the message signal wave is positive and the modulated signal is also positive, the output will be multiplication of both. If the amplitude of the modulating signal is negative at the time of the pulse signal, the sum of both signals will be the difference between the two.

The combined signal is then sampled with the approximated amplitude of signals. The sampled signals are then quantized into a digital value in terms of 0s and 1s so that the digital values of each sample can be calculated. Those streams of quantized signals measure the value of the output PAM signal.

Pulse Code Modulation (PCM)

PCM is a form of pulse modulation. Like PAM, PCM is also accomplished in three different steps. Before those three steps, the input signal (message signal) is passed through

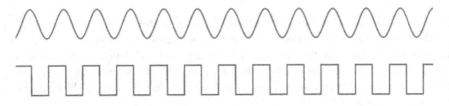

FIGURE 1.10   Pulse amplitude modulation technique.

a low-pass filter so that the input signal is clean and separated from disturbances of high frequencies properly. Three steps of PCM are listed as follows:[29]

- Sampling of the input signal
- Quantization of the input signal
- Encoding of the quantized signals

In PCM, the sampling is done through two different schemes, which are given as follows:

- Differential pulse code modulation (DPCM)
- Adaptive differential pulse code modulation (ADPCM)

Pulsewidth Modulation (PWM)

PWM is the representation of the amplitude of the input analog signal with the longer duration of signal showing the "ON" state, while the rest of the time when the input signal travels through the negative cycle, the digital switch remains off. This is commonly used in electronic controlled power supplies and other communication in machines.[30]

The pulsewidths are variable for maintaining the transistor at the output on for a longer period of time when the input signal has a high amplitude. And similarly, when the amplitude of the input signal goes low, the output transistor remains off to just indicate that the amplitude of the signal has gone low. This technique is extensively used in saving power in electronic circuits to increase the efficiency of the circuits.

Amplitude Shift Keying (ASK)

ASK is a type of analog modulation based on the fundamental amplitude modulation. Amplitude modulation uses different types of keying commonly known as the types of amplitude modulation. In an ASK scheme, the output of a high-frequency carrier is turned on to generate continuously high frequency when the input digital signal is high. The output frequency generator generates no frequency at all when the input digital signal is a zero-value or low-input signal.

Frequency Shift Keying (FSK)

FSK is one of the useful techniques deployed in digital modulation systems. This is a type of FM. In this type of modulation, the output carrier signal of the system varies in terms of frequency change when the value of the input digital signal changes. If the value of the input digital signal is high or (1), the output frequency of the carrier frequency increases. Meanwhile, when the value of the input digital signal is low or zero (0), the frequency of the output carrier goes low.[31]

FSK uses two types of mechanisms, which are listed in the following:

- Synchronous FSK
- Asynchronous FSK

Phase Shift Keying (PSK)

PSK is a type of digital modulation. The carrier band shifts its phase of the frequency by 180 degrees when the input signal changes the state from zero to one or from one to zero. The frequency of the carrier band is constant, but the state of the input signal is used to change the phase of the frequency of the output carrier signal.

This scheme of digital modulation is extensively used in wireless local area network (WLAN), Bluetooth, radio-frequency identification (RFID), and similar kinds of technologies to increase the efficiency and throughput of the wireless link. Different types of schemes of phase shifting are used in PSK modulation. A few major PSK schemes are listed in the following:[32]

- Binary phase shift keying (BPSK)
- Quadrature phase shift keying (QPSK)
- Offset quadrature phase shift keying
- Differential phase shift keying

Quadrature Amplitude Modulation (QAM)

This is one of the most commonly used types of modulation in modern radio telecommunication systems. This type of digital modulation is the combination of the two signals, which are sent to the receiver by adjusting those signals by separating the phases of the signals along with the amplitudes of the carriers. The combination of the amplitude and phase is known as the symbol, which is a bunch of information coded into QAM modulation and sent out to the receiver channel.[33]

The symbols of the information vary. A few examples include 16-QAM, 32-QAM, 64-QAM, and others. This modulation is used to enhance the capacity of wireless networks by encoding multiple signals on one channel.

MODES OF TRANSMISSION IN COMMUNICATION

Telecommunication systems use different modes of transmission in communication. The classification of the transmission modes is done on the basis of the capabilities of the transmission system to support directions of the communication. The modes of transmission used in communication systems are listed in the following:[34]

- Simplex mode of communication
- Half-duplex mode of communication
- Full-duplex mode of communication

As we know, the telecommunication system has evolved through different phases. The capabilities and design of the communication systems in the old days were not as mature and efficient as they are today. They became efficient with the passage of time and advent of the newer technologies. Let us talk about the three major modes of communication.

Simplex Communication

The simplex is a mode of communication channel in which the information is sent out in just one direction. The response from the other direction cannot be sent. The definition of the simplex mode of communication channel is defined by the ITU.

Examples of the simplex mode of communication channel include one-way communication through radio stations, television, and others. In simple words, we can say the simplex mode of communication is based on a monologue system of communication. This mode of transmission channel is considered the poorest one in terms of efficiency.

Half-Duplex Communication

The half-duplex mode of communication is a two-directional channel of communication but one direction of communication works at one time. This is a comparatively better mode of communication channel than the simplex mode of communication channel.

An example of half-duplex mode is walkie-talkie. You can send voice at one time and receive the communication at the other time. This type of communication is known as half-duplex mode of communication.

Full-Duplex Communication

Full-duplex mode of communication refers to the communication channel, which is capable of receiving and sending the communication signals at the same time (or, simultaneously). With this communication channel, both the sender and the receiver can talk and listen to the communication over the channel (at the same time).

An example of full-duplex mode of communication channel is telephone, and all modern interactive and responsive applications that can talk and listen to the communication (from the other side) simultaneously. This is the best performing mode of communication channel used in modern telecommunication systems worldwide.

TYPES OF MULTIPLEXING

Multiplexing is an electronic process in which multiple signals of different frequencies, wavelength, and amplitude are combined to form one single combined signal for transmission over the wireless or other wired medium. The electronic devices that perform the multiplexing process in the communication systems are known as multiplexers. The multiplexers are very common in almost all types of communication signal transmitter and receiver devices.

When multiple signals are combined to form a composite signal for transmission and transmitted over any specific medium, we will also need to handle and process that particular composite signal to fragment it into its original components of input signals. The process of sorting out the input components of the signals in a composite signal received at the electronic receiver side is known as demultiplexing. The demultiplexing process is exactly the opposite of the multiplexing process done at the transmitter end.

The electronic device that separates the input signals from a composite signal received by an electronic receiver is known as a demultiplexer. In real-world transmission systems, the multiplexing and demultiplexing functions are combined in a circuit, commonly known as a multiplexer.

There are two major types of multiplexing in electronic data communication systems. They are listed as follows:[35]

- Analog multiplexing
- Digital multiplexing

These basic types of multiplexing techniques are further divided into other types of multiplexing techniques. Let us discuss analog and digital multiplexing separately along with their subtypes, respectively.

ANALOG MULTIPLEXING

Analog multiplexing is the process of mixing the analog signals with respect to wavelengths and frequencies of the input signals to form a composite signal for transmission. Analog multiplexing uses different types of techniques. Two of those techniques are listed in the following:

- Frequency-division multiplexing (FDM)
- Wavelength-division multiplexing (WDM)

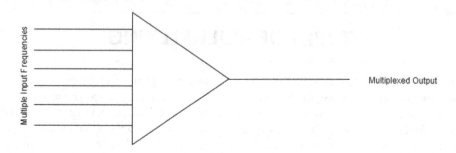

FIGURE 1.11 Schematic diagram of a multiplexer.

Frequency-Division Multiplexing (FDM)

The FDM is a type of technique of multiplexing, in which signals of different frequencies are combined into a multiplexed signal to transmit over a certain medium such as cable, air, etc. The combined signal consisting of multiple frequency channels transmits data of multiple separate channels combined and is transmitted over a medium.

An example of FDM is the transmission of FDM signals of multiple TV channels over a single coaxial cable. A few years back, analog multiplexers were extensively used in the TV cables. But in recent days, the old FDM-based TV cables are being replaced with digital multiplexing to use the cables more efficiently.[36]

The schematic diagram of a multiplexing process is shown in Figure 1.11. In this figure, multiple channels of frequencies are inputted and multiplexed into a single composite signal consisting of those input channels of frequencies ready for transmission over a coaxial cable.

Wavelength-Division Multiplexing (WDM)

WDM is almost a variant of frequency multiplexing in analog electronic communication systems. In this type of multiplexing, multiple input signals of different wavelengths are multiplexed into a composite signal that consists of multiple wavelengths in the output signal. This technique is extensively used in light signals to transmit the data over fiber optic systems.

DIGITAL MULTIPLEXING

The multiplexing of multiple discrete digital signals in the form of ones and zeros in such a way that multiple signals travel together to the destination is known as digital multiplexing. As we know, the digital signals are based on only ones and zeros without any other differentiation like in analog signals such as frequency, wavelength, amplitude, phase, and other factors.

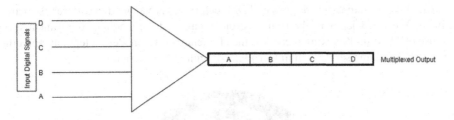

FIGURE 1.12   Schematic diagram of digital multiplexing.

A digital signal consists of zeros (0s) and ones (1s), irrespective of the back-end communication protocols used in the circuits and transmission mediums; so, the types of digital multiplexing are classified into just one major category, which is known as TDM. Numerous other techniques are used in the TDM, but the basic type of digital multiplexing is just one, which is TDM.

The schematic diagram of digital multiplexing is shown in Figure 1.12. In this figure, four digital signals are shown as the input sources of signals named A, B, C, and D. The input data of all those digital signals are based on zeros and ones. The output signal is divided into timeslots A, B, C, and D. In each time slot, the input data (zeros and ones) from each channel are transmitted. This process of transmitting signals from each input channel in a timeslot is repeated to carry out the transmission of data over the specified channel.

TDM offers additional capabilities as compared to analog FDM. A few of them are listed in the following:[37]

- It is more flexible than FDM
- Efficient use of the entire available bandwidth of the channel
- Very simple and easy electronic circuitry
- Lesser cross-talk interference

Time-Division Multiplexing (TDM)

In the TDM technique, the transmission time of the signal is divided into multiple slots, which are allocated for every input digital signal of a communication system. The output signal of multiple input digital signals is governed by a certain rule of communication to produce a combined digital signal, and at the receiving end, the same digital signal is separated into multiple output signals based on a certain mechanism of communication.

For the mechanism to govern this entire process of combining, multiple similar types of signal communication protocols are used. The TDM is further divided into the following subtypes of multiplexing schemes:[38]

- Synchronous TDM
- Asynchronous TDM
- Statistical TDM
- Interleaving TDM

All of the aforementioned types of TDM techniques are applied in different scenarios to increase the efficiency of the transmission signals and the media of the transmission. The use of TDM multiplexing increased the efficiency of our modern telephony and other data communication services in modern telecommunication systems.

Sample Questions and Answers for Chapter 1

Q1. What is an electromagnetic wave? What are the types of it?

A1. Electromagnetic waves are a wide range of electromagnetic frequencies that can travel in the air as well as in the vacuum. Electromagnetic waves are created by a variation of the electric field, which creates a varying magnetic field. The creation of such waves associated with the electric and magnetic field is known as an electromagnetic field. The electromagnetic waves are classified into two domains, which are also interlinked with each other. Those two domains are wavelength and frequency. Both frequency and wavelengths are also used to classify electromagnetic waves.

Q1. What are the major modes of transmission used in communication systems?

A2. The modes of transmission used in communication systems are listed in the following:
- Simplex mode of communication
- Half-duplex mode of communication
- Full-duplex mode of communication

Q3. What is phase modulation?

A3. Phase modulation is a type of angle modulation in analog modulation. The phase modulation technique is also used in digital modulation technology. Phase modulation is a type of modulation technique in which the frequency of the carrier signal is changed in proportion with the change in the phase of the message signal.

Q4. What is multiplexing?

A4. Multiplexing is an electronic process in which multiple signals of different frequencies, wavelengths, and amplitudes are combined to form one single combined signal for transmission over the wireless or other wired medium.

Q5. Name the four major types of TDM.

A5. Four major types of TDM are:
- Synchronous TDM
- Asynchronous TDM
- Statistical TDM
- Interleaving TDM

Introduction to Analog Communication Protocols

2

TELEGRAPH PROTOCOLS

Different types of telegraph systems based on electric current were developed from the early decades of the 19th century. All those systems would use different communication protocols to transmit information from one place to another. The first recognized

DOI: 10.1201/9781003300908-3

telegraph communication protocol was developed by Samuel Morse in 1838. The communication code was named Morse code after the name of the inventor of the code.

Earlier than the invention of Morse code, some other types of codes were in use for local communication. They are not considered as the most important ones as the Morse Code. Before the commercial use of Morse code, different types of optical codes were also in use for transmitting the information from one place to another.

It is very important to note that the early-age telegraphy systems were manual communication systems. With the passage of time and improvements in the techniques, the manual systems were transformed into automatic telegraph systems. In other words, we can divide the telegraphy systems into the following two broader categories.[39]

- Manual telegraphy systems
- Automatic telegraphy systems

The manual telegraphic/telegraphy systems would use the manual telegraph codes, and automatic systems would use the automatic communication codes or protocols. The most commonly used manual telegraph communication codes can be classified into two categories as listed in the following:

- Optical codes
- Electrical codes

There are numerous types of codes that are used in both of the aforementioned categories, i.e., electrical codes and optical codes. Those codes will be discussed one by one in the following sections. On the other hand, the automatic telegraph codes are further divided into two major types as mentioned in the following list:

- Baudot code
- Murry code

Let us know about different codes under the optical, electrical, and modern automatic codes separately.

OPTICAL TELEGRAPHY CODES

Brilliant French scientist Claude Chappe and Swedish researcher Abraham Edelcrantz are considered as the pioneers of developing optical telegraph codes.[40] Both of them developed communication codes, which are named after the names of the inventors of those codes. The major codes used in optical telegraph systems are listed in the following:

- Chappe code
- Wigwag code
- Edelcrantz code

To have a deeper perspective on the major optical telegraph communication protocols, here, we define and explain them at length.

Chappe Code

The Chappe code was first used in an optical telegraph system between two cities: Paris and Lille in France. A chain of towers constituted the optical telegraph system between the two cities in 1793 (as a pilot project), which was put into the service a year later in 1794. This system used Chappe code for the newly installed optical communication system. In this code, the arms or the code structure of the system would be moved by 45 degrees to make eight different positions on the semaphore tower. The semaphore flag positions are shown in the schematic diagram in Figure 2.1.

The Chappe code consisted of 8×4×8=256 code space. Due to some problems in code recognition, the one position from the code space was not used for communication. Thus, the total code space used for the Chappe code was 7×4×7=196. The transmission of the message through the Chappe code was carried out with the help of a code book. The code book consisted of a large number of phrases and words. The Chappe code works on the principle of semaphore flag position in which every flag or code structure was allowed to move 45 degrees.

The modifications and enhancement in the code to make semaphore communication more efficient and useful continued for many years. The new editions of the code book were released after a few years of study and enhancement recommendations.

FIGURE 2.1   Schematic diagram of semaphore flag (Pixabay).

Wigwag Code

The Wigwag code is a type of signaling based on the position of flags held in hands at different angles and positions. Every position was coded to provide meaningful information or action. It was developed extensively for operations in the military fields. Before the advent of new semaphore codes, it was extensively used in optical telegraph systems.

The Wigwag code[41] was developed by Albert J. Myer, who was a military surgeon. He developed the code for army field operations before the civil war in America. In this system, one flag was used to move back and forth to dispatch a message to the person at a distant location. This system was not possible to be implemented into the communication system at a national scale, but it was the foundation for future communication systems in optical telegraphy.

In this coding system, different sizes of flags were used to increase the visibility and effectiveness of the signals. For this, the left, right, and front positions of the flag in hand would indicate the numeric values 1, 2, and 3, respectively. Later on, the flag was replaced with a special lamp powered by kerosene oil to signal it in the nighttime. This system remained in use till the end of the 18th century in different units of the Army in the USA.

Edelcrantz Code

Just after the Chappe code was brought into operations, the Edelcrantz code was introduced in Sweden in 1794 with an experimental optical telegraphy system based on three stations covering a distance of about 12 kilometers. This system was established between Royal Castle Stockholm and Drottningholm Castle in Sweden.

This code was also based on the indicator arms, an almost similar technique to that was used in the Chappe code in France, but the code space and indicators were defined in a different way. The Edelcrantz code consisted of ten iron shutters. The first three shutters of this code were used to define the three-digit octal numbers. The last shutter of this coding system was used for special purpose indication.

The Edelcrantz code provided the code space of 1024 code points. Those code points indicated different words, numbers, phrases, and other control signals. The setting of all shutters at one single time would mean the start of the signal through semaphore. The updated design was also introduced by Abraham Edelcrantz in 1809 in which different design modifications and efficiency factors were introduced.[42]

In the Edelcrantz code, major scientific considerations were taken into account to keep human errors at a low level. The design of the semaphore considered the subtending angle measuring 4 minutes of arc, which is much better to cover subtending angle distortion in the view of the arms and shutters of the semaphore systems. The sizes of the shutters were also designed as per the available scientific research works and findings to make it more accurate. The first operational line for official use was deployed and started in January 1795 between Vaxholm and Stockholm.

ELECTRICAL TELEGRAPHY CODES

The discovery of electricity paved the way for communication based on electric charges, currents, and voltages. The use of electric current in the telegraph system was introduced by Samuel Morse by developing a proper code for communication through electrical pulses over the copper cables.

Before the Morse code, the use of electric signals could also be traced in different countries in Europe and North America, but it was not properly patented or officially announced with proper commercial use of those codes.

Let us now talk about a few most common electric telegraphy codes that were extensively used in telegraphy services in the world.

Morse Code

Morse code is the first revolutionary code that was extensively used for over one and half century across all continents of the globe. Morse code was invented by Samuel Finley Breese Morse (Samuel F.B. Morse), who was a scientist and painter. He was born in Charleston town in the state of Massachusetts, USA. Later on, he moved to New York, where he spent his remaining life and also died there.[43] Morse code was invented in 1838. The role of Alfred Lewis Vail cannot be discounted because he was an assistant as well as the business partner to Samuel Morse in developing, improving, and implementing the code for business purposes. Alfred Lewis Vail also improved the Morse code by some modifications later on.

Morse code consists of dots, dashes, and spaces to code numerals, alphabet letters, and punctuation marks. Different letters, numbers, and punctuation marks were coded with different sizes of dashes, spaces, and dots. The variable sizes of dashes were generated by sending the electrical current for variable times at the point of sending the signals, which would produce voltage at the receiving end through armature (in electrical engineering, an armature is the component of an electric machine that carries alternating current) and the coil would drag the printing head on the paper to mark spaces, dots, and dashes of different patterns. The definition of those spaces, dots, and dashes was defined under the Morse code of electrical telegraphy. The Morse code encoded the numbers from 0 to 9 and alphabetic letters from A to Z. The total number of initial communication symbols and letters that were encoded through the Morse code was 36. Later on, different signs and special characters were defined in different versions of the Morse codes.

Initially, the Morse code was designed from the perspective of the English language that is shown in Figure 2.2, which shows the code for 36 characters (including numerals or numerical characters). So, it was modified for international use, when the electric telegraphy went into Europe and other parts of the world. The newly revised version of the Morse code is also known as the International and Continental Morse Code, which was able to accommodate non-English languages.

A .-	J .---	S ...	1 .----
B -...	K -.-	T -	2 ..---
C -.-.	L .-..	U ..-	3 ...--
D -..	M --	V ...-	4 ....-
E .	N -.	W .--	5 .....
F ..-.	O ---	X -..-	6 -....
G --.	P .--.	Y -.--	7 --...
H	Q --.-	Z --..	8 ---..
I ..	R .-.		9 ----.
			O -----

FIGURE 2.2   Morse code system.

The newly developed continental Morse code would use the combinations of dots, spaces, and dashes. The equal lengths of dashes were standardized for use. The use of equal size of dashes made the understanding of the message easier. The combination of multiple dashes, spaces, and dots was also possible to use.

The Morse code was extensively used across the world, especially in different wars such as the Korean war, World War II, Vietnam war, and many other wars. The use of Morse code continued till the advent of modern communication systems powered by computer networks in the early 1990s. It ruled the majority of the land communication for a long period of time. The importance of the Morse code did not diminish despite the extensive use of telephone service after its invention in 1876–77. The combination of telephone and telegraph was a great communication solution for government record keeping through telegraph and informal communication through telephone.[44]

Cooke and Wheatstone Code

The Cooke and Wheatstone code is another code for electrical telegraphy. Initially, Morse code and Cooke and Wheatstone code were considered as the competitors with each other. Both codes had separate devices for transmitting and receiving signals. But, the extensive use, simplicity, and flexibility of scaling up the communication code helped the Morse code completely vanish the Cooke and Wheatstone code within a few years of competition.

Cooke and Wheatstone code is named after the names of the inventors. William Fothergill Cooke and Charles Wheatstone were two English inventors and scientists who developed an electric telegraph system, which consisted of an electrical bridge consisting of five needles. When a certain pattern of current would pass through the circuit, the needle would move to a character, which indicated the part of a message.

Wheatstone bridge electric telegraph was much simpler and easier to use, but the use of Morse code was so extensive across the world that this system could not get enough grounds. The entire construction of the Cooke and Wheatstone bridge consisted of five needles to point the character such as alphabetic characters and a grid of 20 characters of

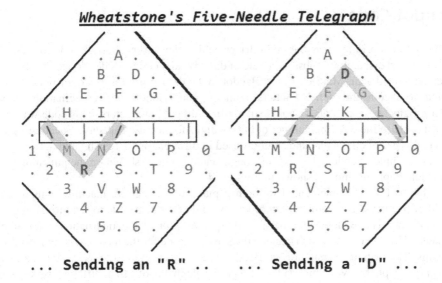

FIGURE 2.3   Cooke and Wheatstone telegraphy code (Flickr).

alphabet, which would constitute a complete assembly of the telegraphy system.[45, 46] The full system of Cooke and Wheatstone telegraphy with five needles is shown in Figure 2.3 (conceptual diagram).

As shown in the figure, two needles were used as switches to complete a circuit for sending R in the matrix of the characters. Similarly, to send D, in the right part of the diagram, one needle would be connected to the row consisting of I, F, and D and the other needle would be connected to the circuit, which has a series of characters, i.e., L, G, and D. Thus, the circuit (to indicate D) from two different paths would get completed and the signal for D would be sent to the receiving machine, which would detect the signal and indicate the D on the screen on the receiver device.

Cooke and Wheatstone electric telegraphy used three variations of the codes. Those codes were based on the one-needle system, three-needle system, and five-needle systems. All of those three systems would use different codes for the transmission of messages. The main problem of this coding system was the fixed assembly, which had very limited scope for scaling up and modifications without a complete overhaul. This is the main reason that the Morse code completely defeated it in the marketplace.

Automatic Telegraph Codes

In the aforementioned sections, the major electrical codes used in the manual telegraphy systems have been discussed in full detail. Let us discuss the most modern automatic telegraph codes, which were used in modern telegraphy. After the advent of teleprinter machines, the need for speed became a major point of interest for scientists to develop new codes that could transmit bigger information rates at a faster speed. In that pursuit, the following automatic codes were developed for achieving a faster telegraphic service.

Baudot Code

Automated teleprinting was the main driver behind developing Baudot code to increase the speed and dispatching substantial rate of data. Baudot code was developed by a French scientist named Jean-Maurice-Émile Baudot in 1870s. He used a five-digit binary code system to encode the alphabet letters, numeric, and symbols. The combination of five binary digits (zeros and ones) would suffice in coding 32 characters. This code was also very flexible to increase the number of binary digits to increase capacity to accommodate different types of characters and symbols used in different non-English languages.[47]

The sample of coded characters through Baudot code is shown on the punching tape of the telegraph machine as shown in Figure 2.4.

This code was implemented through a piano-like telegraph machine, which consisted of five keys. The combination of keys would generate a code to send to the receiving machine. The Baudot code became the predecessor of the International Telegraph Alphabet Version ITA-2, which became the standard code for the then modern teleprinting systems. The primary version of the Baudot code is also referred to as the International Telegraph Alphabet Version 1 (ITA-1). The development of other different versions of codes based on seven-binary digits, eight-binary digits, and other versions of the digital codes was also driven by the ideas of Baudot. The unit of transmission rate of information is baud, which is also named after the scientific work done by Baudot.

Murray Code

Murray code is a modified version of the Baudot code. This code was developed in 1901. This was needed after the invention of the typewriter-like machine by Donald Murray. He was born in New Zealand and worked in Australia, London (UK), New York (USA), and other countries in Europe for a long time. He is also known as the inventor of the telegraphic typewriter.

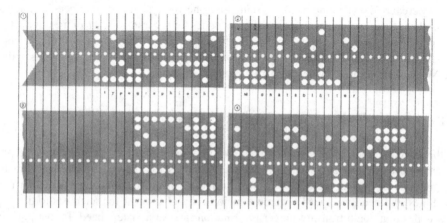

FIGURE 2.4   Baudot code (Flickr).

FIGURE 2.5 Teleprinting machine using Baudot–Murray code (Flickr).

The Murray code was the modified version of the Baudot code. Later on, a combined code developed for teleprinting was known as Baudot–Murray code. That code was also recognized as the International Teleprinter Code (ITC). The Murray code used a series of electrical pulses and no-pulses, hole punches, and no-hole punch spaces based on the five-digit code. The Murray code also incorporated different operations such as addition, subtraction, and other components into the codes. A typical teleprinting machine is shown in Figure 2.5.

The most frequently used letters were coded in such a way that they required the least number of hole punches in the code. The letters that are the least used in the communication like X, K, Q, and V were coded with the four-hole punches, the maximum number of holes in the coding system. Murray code was heavily used by Western Union for a very long time up to the 1950s. The Murray code was also used in many countries and across the world until the ASCII codes were developed later on for standard and extended communication systems.[48]

ASCII Code

American Standard Code for Information Exchange, precisely ASCII is an advanced automatic code for computer-based printing. This code was developed by a committee of the American Standard Association (ASA) in 1963. The name of the committee was X3 Committee. Primarily, the committee developed the ASCII code based on seven (7) binary digits, which could cater to 128 characters and symbols. Those characters consist of control functions as well as printable characters.[49] The first publication of the ASCII standard appeared under the name ASA X3.4–1963. Numerous revisions of this document were regularly published subsequently. The 7-bit ASCII code is shown in Figure 2.6.

As we know, there are many function keys on modern computers that are designated for certain functions. To convey the signal of any kind of control function on the computer, you

Dec	Hx	Oct	Char		Dec	Hx	Oct	Html	Chr	Dec	Hx	Oct	Html	Chr	Dec	Hx	Oct	Html	Chr
0	0	000	NUL	(null)	32	20	040	&#32;	Space	64	40	100	&#64;	@	96	60	140	&#96;	`
1	1	001	SOH	(start of heading)	33	21	041	&#33;	!	65	41	101	&#65;	A	97	61	141	&#97;	a
2	2	002	STX	(start of text)	34	22	042	&#34;	"	66	42	102	&#66;	B	98	62	142	&#98;	b
3	3	003	ETX	(end of text)	35	23	043	&#35;	#	67	43	103	&#67;	C	99	63	143	&#99;	c
4	4	004	EOT	(end of transmission)	36	24	044	&#36;	$	68	44	104	&#68;	D	100	64	144	&#100;	d
5	5	005	ENQ	(enquiry)	37	25	045	&#37;	%	69	45	105	&#69;	E	101	65	145	&#101;	e
6	6	006	ACK	(acknowledge)	38	26	046	&#38;	&	70	46	106	&#70;	F	102	66	146	&#102;	f
7	7	007	BEL	(bell)	39	27	047	&#39;	'	71	47	107	&#71;	G	103	67	147	&#103;	g
8	8	010	BS	(backspace)	40	28	050	&#40;	(	72	48	110	&#72;	H	104	68	150	&#104;	h
9	9	011	TAB	(horizontal tab)	41	29	051	&#41;	)	73	49	111	&#73;	I	105	69	151	&#105;	i
10	A	012	LF	(NL line feed, new line)	42	2A	052	&#42;	*	74	4A	112	&#74;	J	106	6A	152	&#106;	j
11	B	013	VT	(vertical tab)	43	2B	053	&#43;	+	75	4B	113	&#75;	K	107	6B	153	&#107;	k
12	C	014	FF	(NP form feed, new page)	44	2C	054	&#44;	,	76	4C	114	&#76;	L	108	6C	154	&#108;	l
13	D	015	CR	(carriage return)	45	2D	055	&#45;	-	77	4D	115	&#77;	M	109	6D	155	&#109;	m
14	E	016	SO	(shift out)	46	2E	056	&#46;	.	78	4E	116	&#78;	N	110	6E	156	&#110;	n
15	F	017	SI	(shift in)	47	2F	057	&#47;	/	79	4F	117	&#79;	O	111	6F	157	&#111;	o
16	10	020	DLE	(data link escape)	48	30	060	&#48;	0	80	50	120	&#80;	P	112	70	160	&#112;	p
17	11	021	DC1	(device control 1)	49	31	061	&#49;	1	81	51	121	&#81;	Q	113	71	161	&#113;	q
18	12	022	DC2	(device control 2)	50	32	062	&#50;	2	82	52	122	&#82;	R	114	72	162	&#114;	r
19	13	023	DC3	(device control 3)	51	33	063	&#51;	3	83	53	123	&#83;	S	115	73	163	&#115;	s
20	14	024	DC4	(device control 4)	52	34	064	&#52;	4	84	54	124	&#84;	T	116	74	164	&#116;	t
21	15	025	NAK	(negative acknowledge)	53	35	065	&#53;	5	85	55	125	&#85;	U	117	75	165	&#117;	u
22	16	026	SYN	(synchronous idle)	54	36	066	&#54;	6	86	56	126	&#86;	V	118	76	166	&#118;	v
23	17	027	ETB	(end of trans. block)	55	37	067	&#55;	7	87	57	127	&#87;	W	119	77	167	&#119;	w
24	18	030	CAN	(cancel)	56	38	070	&#56;	8	88	58	130	&#88;	X	120	78	170	&#120;	x
25	19	031	EM	(end of medium)	57	39	071	&#57;	9	89	59	131	&#89;	Y	121	79	171	&#121;	y
26	1A	032	SUB	(substitute)	58	3A	072	&#58;	:	90	5A	132	&#90;	Z	122	7A	172	&#122;	z
27	1B	033	ESC	(escape)	59	3B	073	&#59;	;	91	5B	133	&#91;	[	123	7B	173	&#123;	{
28	1C	034	FS	(file separator)	60	3C	074	&#60;	<	92	5C	134	&#92;	\	124	7C	174	&#124;		
29	1D	035	GS	(group separator)	61	3D	075	&#61;	=	93	5D	135	&#93;	]	125	7D	175	&#125;	}
30	1E	036	RS	(record separator)	62	3E	076	&#62;	>	94	5E	136	&#94;	^	126	7E	176	&#126;	~
31	1F	037	US	(unit separator)	63	3F	077	&#63;	?	95	5F	137	&#95;	_	127	7F	177	&#127;	DEL

FIGURE 2.6   ASCII code chart (Flickr).

need a code in the binary language, which is understood by electronic machines. With the advent of modern computer systems, ASCII code with seven digits was not sufficient to cater all required control functions and printable characters. Therefore, the number of digits in the code was revised with the 8-bit code. Thus, the total number of codes catered by the new version of ASCII code became 256 characters, which is sufficient for modern computing systems.

The 8-bit binary system ASCII code was named Extended ASCII code. This explicitly indicates that the base of the code is the same and just one bit was increased. The characters and controls defined in the 7-bit binary ASCII code remained intact in the extended code, and more characters and control functions were included in the new ASCII code.

The ASCII code is divided into different segments. The first 32 positions (from 0 to 31) of the code define different control functions such as backspace, end of text, start of text, null character, escape, and other control of functions used in modern computers. The next 96 positions (from 32 to 127) encode different printable characters such as alphabet characters in both capital and small formats, numbers, and special signs and characters. The third segment of ASCII code, which is also the extended part of the code, encodes different printable characters and Latin characters.[50]

ANALOG TELEPHONY COMMUNICATION PROTOCOLS

Telephony system was one of the biggest revolutions in the field of telecommunication after electrical telegraph systems. Telephone system evolved through the work of different scientists from UK, Italy, and USA, but the patent was achieved by the US scientist Alexander Graham Bell in 1876. Bell was born in Edinburg, Scotland. But later on, he moved to the USA. The basic rule of communication over telephone was based on the modulation and transmission of the analog signals over a copper wire. Those copper wires were already in use for the telegraph systems. Telephone system is another form of electrical signal communication.[51]

The basic protocols used in the analog telephone system were about the signaling systems. The transmission of signals was based on the modulation of the signals over the carrier frequency, which has already been discussed before. It is very important to note that telephone communication uses signaling systems for four types of purposes for establishing, monitoring, and disconnecting the phone successfully. The four purposes of telephone communication protocols are listed as follows:

- Alerting for incoming call
- Addressing destination
- Supervising idle telephone lines
- Informing the status through response

Initially, the telephone system evolved from single-line phones to multiple lines and telephone exchanges or offices. So, the types of signaling can also be divided into two categories. The local loops are connected within one office premises, while the interoffice connections of telephones are located in two different locations connected through trunk lines. So, we can divide telephone signaling into the following categories:[52]

- Call-Number Dialing Protocols
- Interoffice Signaling Protocols

Let us now explore the types of protocols used in the aforementioned major categories of telephone protocols.

CALL-NUMBER DIALING PROTOCOLS

The dialing phone number of a line in the beginning was not required because every physical circuit was assigned a dedicated number and the exchange operator would connect the desired lines. But, the dialing protocols for call-number were required very soon. The wire-loop current sending signaling to alert incoming telephone systems was soon replaced by different types of signaling systems or protocols as mentioned in the following.

Rotary Dialing

The rotary dialing protocol is also known as the pulse dialing system. The rotary dialing was invented by Almon Brown Strowger, an American inventor. He invented rotary

FIGURE 2.7   Rotary dialing phone (Flickr).

dialing in his electromechanical telephone exchange system in 1891. The switch would use rotary dialing to send signals to the central exchange.[53]

In the rotary dialing, a finger wheel was used. There were 10 slots on the wheel named 1, 2, 3, 4, 5, 6, 7, 8, 9, and 0. The zero would be coded as 10 pulses of signals. Each number of the figure wheel was located at such a position that the smooth revolving of the wheel by fixing a figure in the slot of the desired number would send the pulses equal to that number, which is selected by inserting the figure in the slot. The figure wheel was coupled with a recoil spring at the back-end. The coil would recoil the wheel back to its original position, where the dialer can choose the other number to dial. This dialing system was a bit difficult and slow to dial. A smooth and precise rotation of the wheel was required to dial the desired number. The rotary dialing phone is shown in Figure 2.7.

This is a type of electromechanical dialing in which the pulses of currents are sent to the exchange. The current pulses were generated by the rotary wheel at a rate of about 10 pulses per second. The pulse rate coding of those pulses was a bit different in different countries, but the basic principle of rotary dialing was the same all over the world.

The other version of rotary dialing was the switch-hook dialing, which is not a standard type of dialing, but the dialing process can be completed through that method by using a tricky technique to tap the telephone hook fast (many times in a second). This unrecognized dialing system is mentioned in the following sub-section under the switch-hook dialing system.

Switch-Hook Dialing

This is another type of dialing, which was not recognized as the commercial dialing protocol to use. The basic principle of this dialing system was the same as rotary or pulse dialing. In fact, we can say it is a bug in pulse dialing. Actually, the *pulse train* is sent to transmit the number to the destination telephone. To clarify here, a pulse wave or pulse train is a kind of non-sinusoidal waveform that includes square waves (duty cycle of 50%) and similarly periodic but asymmetrical waves (duty cycles other than 50%). The pulse train is based on switching the hook certain times such as one time for dialing one, two times for dialing two, three times for dialing three, and so on.

So, when a hook is tapped three times fast, the system will send a pulse of 3 to the destination. In such a way, a person can dial the entire telephone number by fast tapping on the hook. This technique was misused in coin phones in the UK and other countries, where dialing pulse was sent through switch-hook dialing without using the coins. So, we can say this was neither a recognized protocol for telephone dialing nor used in commercial applications.

Dual-Tone Multi-Frequency (DTMF)

Dual-Tone Multi-Frequency (DTMF) is a type of dialing. It is also known as the "Push button dialing". In this type of telephone number dialing system, a grid of buttons is used for transmitting a pair of frequencies, which denotes a digit ranging from 0 through 9 and two signs "*" and "#".

The grid of buttons consists of three columns and four rows as shown in Figure 2.8.

A substantial number of research studies and practical experiments were conducted for push-button dialing to check its efficiency, easiness, and robustness of this type of dialing in the 1950s. It was found in many research studies that this dialing was much more efficient than its predecessor rotary dialing. In 1963, the US telecom incumbent AT&T

FIGURE 2.8   Push-button dialing pad (Flickr).

launched the commercial use in the country followed by acceptance by all countries in the world. This system of dialing is known as MF4 in the United Kingdom. The International Telecommunication Union – Telecommunication (ITU-T) standardized the DTMF dialing under the Q.23 recommendations.

As the name of DTMF indicates, it uses two frequencies for every push of the button on the grid of the dialing pad. In the dialing pad grid of buttons, each row of the pad is assigned one frequency, which is different from the frequency of the other rows, while each column of the pad is assigned different frequencies. When the button on the dialing pad is pushed, both the frequencies – corresponding row and column frequencies – are generated through an internal frequency generator and sent over the wire to the exchange of telephone system.

In this dialing system, the frequencies used in the rows from top to bottom are 697, 770, 852, and 941 Hz, respectively. The frequencies assigned to each column of the dialing pad grid of the touch-tone dialing system of the telephone from left to right are 1,209, 1,336, and 1,477 Hz, respectively. When any button on the dialing pad is pushed, the two levers connected to that particular button activates the frequency generator to generate two corresponding frequencies associated with the row and column of that particular button in the dialing pad matrix.[54]

Those two frequencies associated with a particular button are superimposed by the mixer of the frequencies and sent out to generate a signal for the destination exchange to dial that particular telephone number, which is dialed with the help of the combination of two frequencies superimposed in one frequency. In many telephone systems, the fourth column is also added in the dialing pad or matrix. That column displays A, B, C, and D. The frequency of the fourth column on the dialing matrix of the telephone is 1,633 Hz. Those characters are used for the control functions in the amateur radio systems in the USA and the other countries of the world.

Multi-Frequency (MF) Signaling Protocol

Multi-frequency (MF) signaling was used in the early telephone signaling protocols. In this form of signaling, the combination of two pairs of frequencies was used to send out the source address information to the destination office. That pair of signals would create sound at the destination office. The MF system would send the signals through the transmission of multiple frequencies for signaling and dialing purposes. This signaling protocol was the predecessor of the DTMF system. MF signaling protocol was used to access the trunk between two telephone offices. This signaling is a type of common channel signaling (CCS), which is an in-band type of telephone signaling extensively used as the local signaling system.

The two versions of this in-band signaling system were common in use. MF compelled the MFC R1 local signaling system, which was used in the USA. The other version of the same signaling system was MFC R2 local signaling system. The MFC R2 signaling system was extensively used for Europe and other countries. Both of those signaling systems will be explained in the subsequent sections separately.

INTEROFFICE SIGNALING PROTOCOLS

The interoffice signaling protocols are the protocols that are used to supervise the telephone line status and establish a telephone call, then supervise the call, and finally end the call after completion. The entire process is handled by a signaling ecosystem consisting of numerous electric and electromagnetic messages and signals. The combination of those messages, signals, and their responses to establish the telephone communication between two speakers is known as the interoffice signaling protocol.

The interoffice signaling protocols can be divided into two major categories in terms of the status of the channels used in the signaling:

- CAS
- CCS

Both of the aforementioned categories of signaling systems have different types of telephone signaling systems, which were commonly used in the past and a few are even used in the present-day digital era of telephony systems. As this chapter is dedicated to the analog telephone system protocols, the major types of CAS will be discussed at length. The digital signaling systems will be discussed in length in the next chapters.

CHANNEL-ASSOCIATED SIGNALING (CAS)

CAS is a type of signaling, which uses the same channels for signaling to control the call establishment, communication, and disconnection that are also used for transmission of voice. This is a generic category of signaling, which is characterized by the use of the same channels for both signaling and voice transmission. There are numerous types of signaling systems that fall under the CAS systems. A few of those CAS signaling systems are listed in the following:[55]

- Loop-Start Signaling Protocol
- Ground-Start Signaling Protocol
- Ear and Mouth (E&M) Signaling Protocol
- MF Compelled R1 Signaling
- MF Compelled R2 Signaling
- Signaling System No. 4
- Signaling System No. 5
- Signaling System No. 6

All of the aforementioned types of signaling protocols use different techniques such as MF, current loops, hook-ups, and others.

Now, we will discuss the aforementioned list of protocols with sufficient details for developing a better understanding of the systems.

Loop-Start Signaling Protocol

Loop-start signaling protocol is a type of protocol used between a local office (exchange) or a private branch exchange (PBX) and the telephone subscriber line. In this signaling system, the state of the telephone line was determined – whether the telephone line is on-hook or off-hook. This signaling system was initially used for checking the status of the subscriber line only. But later on, an additional capability to control the disconnection and supervision of the line communication was also added through the Kewlstart software for the advanced digital telephone systems.

In this protocol, during the on-hook state of the line, the PBX or local office maintains a voltage of −48 V in the ringing line or conductor. This voltage is commonly known as the nominal voltage. As the hook of the telephone is turned to "OFF" state or hook-off state, the loop is completed between the tip conductor and the ring conductor, which initiates the dial-tone to signal the user to dial-in the desired number that he/she wants to talk to. This mechanism is called loop-start signaling. [56]

Similarly, for alerting for the incoming call, the exchange superimposes the alternating current based on alternating voltages ranging from 40 to 90 V. The frequency of that alternating voltage is about 20 Hz. This superimposed voltage will ring the bell in the telephone set.

Ground-Start Signaling Protocol

Ground-start signaling protocol is also known as the successor of loop-start signaling. There were certain drawbacks of the loop-start signaling system, which were solved with the help of the ground-start signaling system. A few of those drawbacks are listed in the following:

- The detection of simultaneous seizure of channel/line was not possible in the loop-start signaling system
- No capability of the protocol to provide telephone exchange to inform for the conditions of the disconnection at the far-end of the call

Those drawbacks were largely solved with the ground-start signaling protocol. In this protocol, the subscriber or the PBX uses the grounding of the ring conductor, which is sensed by the local office. In the idle state of the line, the central exchange office (CO) sets the voltage of the ring conductor to −48 V with respect to the tip conductor. The PBX that initiates the trunk checking for starting a call will ground the ring conductor, which is sensed by the local office, and also grounds the tip conductor to ground. Thus, the connection loop is completed and the dial tone is passed. [57]

Ear and Mouth (E&M) Signaling Protocol

This is another important type of analog telephone signaling protocol, which was initially developed for signaling between two private automatic branch exchanges (PABXs) or central office over analog trunks. Later on, this signaling system was also enhanced and extended for the advanced uses in even today's modern IP telephony to detect the off-hook conditions of the telephone trunk or line. This signaling protocol is also referred to as "RecEive and TransMit" as well as Electrical ground (E) and Magnet (M).

In this signaling protocol, two types of wires known as E-lead and M-lead are used for sending direct currents as signaling for detecting and seizing the analog trunk for call establishing. Two sides of the interfaces are defined in the standards developed by the Bell Labs named trunk circuit and signaling unit. Both those interfaces use battery and ground as the helping entities to establish the seizure of the trunk.

In idle connection, the E-lead is open and the M-Lead is connected to the ground. The off-hook condition is sent to the office to indicate the off-hook condition of the trunk by PABX. The E-lead is connected to the ground from the exchange to indicate the off-hook condition. The dialing tone is initiated, and the dialing signals are received.

Three types of start dialing supervision signaling techniques are used in the E&M signaling protocol. They are[58]:

- Immediate start dialing
- Wink start dialing
- Delay dialing

The E&M signaling system uses five types of interfaces as listed in the following:

- E&M Type I – North American standard
- E&M Type II – Two-node back-to-back connection type
- E&M Type III – Modern telephony system
- E&M Type IV – Use signal battery and signal ground
- E&M Type V – Two signaling-node back-to-back connection support

Multi-Frequency Compelled R1 Signaling

MF Compelled MFC R1 frequency is an analog T1 line signaling system. It is also used for digital signaling systems, but mainly it was developed for analog signaling. It is used in North America. So, it is always known as MFC American version. Its counterpart for E1 signaling used in Europe and other countries of the world is known as the European version of MFC R2. The European version R2 will be discussed in the next topic.

MFC R1 is an ITU-T signaling standard defined under Q.310 and Q.332 recommendations. This signaling system consists of two parts of signaling as listed in the following[59]:

- Line signaling
- Register signaling

The line signaling is based on the line status and supervisory signals. The signaling uses a tone signal through a frequency of 2,600 Hz, which is used for analog trunk signaling. As mentioned earlier, this signaling system is also used for digital trunks. In that case, ABCD signals are used for line signaling. For example, the presence of the tone on the trunk is indicated by the zero value of A, i.e., (A=0). For the absence of the tone on the trunk, A-1 signal is used.

The register signals are used for addressing purposes. The combination of two frequencies is used for sending one signal. It works like the pulse dialing system does, but the frequencies and matrix are different. In the pulse dialing system, a matrix of three columns and four rows was used, but in this case, the matrix is different. An upper band of frequencies consists of five frequencies in a row starting from 900 Hz and ending at 1,700 Hz. Each frequency in the series of upper frequencies is 200 Hz larger than the previous one. So, the series of the upper band frequency series consists of 900, 1,100, 1,300, 1,500, and 1,700 Hz. Similarly, the lower frequency band also consists of five different frequencies in a series forming a column starting from 700 Hz, with 200 Hz increment from top to bottom. So, the frequencies used in the MFC R1 protocol for the register signaling for the lower band frequency series consist of 700, 900, 1,100, 1,300, and 1,500 Hz.

The combination of frequencies that are used for different digits and other functions are listed in Table 2.1.

In MFC R1 signaling, digit 1 is sent through the combination of frequencies 900 Hz from the upper frequency band and 700 Hz from the lower frequency band. Similarly, other digits are also sent out as per the combination shown in Table 2.1. The special signals used in the North American system are also shown in the table.

A few important features of the MFC R1 signaling system as a whole are listed in the following:

- It is an in-band frequency signaling system
- A type of CAS system
- Offers larger time tolerance between dial tone and dialing process
- Continuous signaling system
- It is developed in C and Assembly languages
- Allows user callable functions
- Lightweight program with little memory usage
- Supports both analog and digital trunk signaling
- Uses only forward register signals

TABLE 2.1   MFC R1 Register Signaling.

LOW FREQUENCIES (HZ)	HIGH FREQUENCIES (HZ)				
	900	1,100	1,300	1,500	1,700
700	Digit 1	Digit 2	Digit 4	Digit 7	KP3P or ST3P
900		Digit 3	Digit 5	Digit 8	KP or STP
1,100			Digit 6	Digit 9	KP
1,300				Digit 0	KP2P or ST2P
1,500					ST

Multi-Frequency Compelled R2 Signaling

MF compelled R2 signaling system or protocol is the European version of MFC R1 used in the North American Countries. This signaling system is also defined by the ITU-T standard organization under the recommendations named Q.400 to Q.490. The MFC R2 signaling has numerous variants used in different countries of the world.[60]

The main features of MF Compelled R2 signaling are listed in the following:

- It is a continuous form of line signaling
- Out-of-band line signaling system
- In-band register signals based on MFs
- End-to-end and compelled register signaling
- A type of CAS signaling
- Link-by-link type of line signaling
- Supports both analog and digital signaling
- A and B bits are used for digital line signals
- Uses both forward and backward registers for signaling

The analog signaling is based on MF signals. The register signals of MFC R2 support two separate groups of high frequencies as well as two separate lower-frequency groups. The first group of higher and lower frequencies forms a set of signals used for the forward register signals. And, the second group of the lower and upper frequencies forms another matrix for backward register signals. The backward register signals are used for the acknowledgment of the forward register signals.

The forward register signal table consists of 15 forward signals, which are a combination of two types of frequencies – upper frequency and lower frequency. In response to the forward register signal, the opposite telephone exchange sends a signal through the backward register signal, which is the combination of upper and lower frequencies but not in the forward register matrix. The backward register signals are generated by the combination of other frequencies. Both forward register and backward register signaling are shown in Table 2.2 and Table 2.3.

The forward register signals are a combination of high frequencies starting from 1,500 Hz with an increment of 120 Hz up to 1,980 Hz as the fifth member of this series

TABLE 2.2   MFC R2 Forward Register Signals.

LOW FREQUENCIES (HZ)	HIGH FREQUENCIES (HZ)				
	1,500	1,620	1,740	1,860	1,980
1,380	Fwd 1	Fwd 2	Fwd 4	Fwd 7	Fwd 11
1,500		Fwd 3	Fwd 5	Fwd 8	Fwd 12
1,620			Fwd 6	Fwd 9	Fwd 13
1,740				Fwd 10	Fwd 14
1,860					Fwd 15

TABLE 2.3   MFC R2 Backward Register Signals.

LOW FREQUENCIES (HZ)	HIGH FREQUENCIES (HZ)				
	1,140	1,020	900	780	660
1,020	Bkwd 1				
900	Bkwd 2	Bkwd 3			
780	Bkwd 4	Bkwd 5	Bkwd 6		
660	Bkwd 7	Bkwd 8	Bkwd 9	Bkwd 10	
540	Bkwd 11	Bkwd 12	Bkwd 13	Bkwd 14	Bkwd 15

of frequencies as shown in Table 2.2. Similarly, the forward register signal's lower frequency starts from 1,380 Hz and increases with 120 Hz to reach 1,860 Hz as the fifth frequency of this series of frequencies as shown in the table. This table caters 15 forward signals used for the register signaling in the R2 protocol of E1 trunk signaling, which is extensively used in the European and other countries except for the North American countries.

The backward register signals are the combination of higher and lower frequencies in the same way as the forward register signals. But, the frequencies in the backward signaling are used differently as shown in Table 2.3. Opposite to the forward register signaling frequencies, in the backward signaling system, the frequencies decrease in both cases, i.e., low-frequency and high-frequency series from left to right.

From left to right in high-frequency series, 120 Hz is decreased from the previous frequency. Similarly, for the low-frequency series, 120 Hz is decreased from top to bottom sequence as shown in Table 2.3. The backward signals are also indicated in the reverse direction as compared to the forward register signals.

As mentioned earlier, MFC R2 is also used in digital signaling. In that case, ABCD bits are used for the digital trunk signaling. The four bits named ABCD can accomplish 16 different combinations in digital signaling due to binary combinations of the bits.

Signaling System No. 3

Signaling No. 3 is a type of voice frequency (VF) signaling as compared to the R1 and R2 signaling systems, which were based on the MF signaling system. The signaling system No. 3 was based on the 1 VF with their digital conversion of just one digit. The VF used in the CCITT No. 3 signaling is 2,280 Hz. The commercial use of this signaling was not launched due to the limited scope and capabilities of this signaling system. So, it could work to handle two states of the register either tone-present or tone-absent. This signaling system was defined and standardized by the Consultative Committee on International Telephony and Telegraphy (CCITT). The CCITT is now known as the ITU. This is the reason why signaling system 3 is also known as CCITT No. 3 signaling system.[61]

The use of CCITT No. 3 signaling was replaced by the use of a broader signaling system known as CCITT No.4, which will be discussed in the next sub-section.

Signaling System No. 4

The signaling system No. 4 is another type of VF-based signaling system of telephony system. It is the successor of CCITT No. 3, which also used the VF frequency as the core signaling mechanism. Signaling No. 3 used just 1VF, while the signaling CCITT No. 4 uses the 2VF frequencies, which allows the system to use two frequencies for inter-register signaling in analog systems and two digits for the digital systems.

Two voice frequencies used in this signaling system include 2,040 and 2,400 Hz. Those frequencies are also referred to as X and Y frequencies, respectively. This system was widely used in Europe and North Africa until the R2 signaling system replaced it in the 1970s.

Signaling System No. 5

Signaling System No. 5 is an international direct distance dialing system used for inter-continent signaling between Europe and North America. This signaling system was defined as a standard in 1964 and was put into operations in 1970 officially. This signaling system is also referred to as CCITT5, CC5, or Atlantic code. The predecessor of this signaling system was CCITT No. 4, and it was succeeded by the CCITT No. 6 system. [62]

Signaling system 5 is based on the use of MF for in-band signaling. It used six frequencies for addressing and controlling functions. Those frequencies included 700, 900, 1,100, 1,300, 1,500, and 1,700 Hz. Those frequencies were also recognized as A, B, C, D, and E, respectively. In this system, a combination of two frequencies is used for a signaling code. The CC5 signaling system codes consist of 15 codes as forward inter-register signals. This system does not support backward inter-register signals. The main features of this signaling system are summarized as follows:

- It is an MF system.
- It uses six frequencies for signaling.
- It supports forward inter-register signaling.
- It does not support backward inter-register signaling.
- The combination of two frequencies is used for a signaling code.
- It is a form of the in-band signaling system.
- The first five frequencies are used for encoding digits.
- Two frequencies of the first five frequencies are used for encoding any digit from 1 to 9 and zero.
- The combination of the last frequency with the first five frequencies (combination of both) is used for five additional control functions.

- The sixth frequency is used for the end of the sequence in combination with other five frequencies.
- This system sends the entire number in a bloc. The number codes are stored in the register before sending out to the international tandem exchange (Gateway).
- The first frequency in combination with the last frequency is known as Keying Prefix (KP) code.
- The last frequency in combination with the last frequency is known as Keying Finish (KF) code.
- It supports eco-compressor capabilities.
- It is highly suitable for satellite communication signaling for international telephony.
- It is suitable for two-way operations.
- The post dialing delay is larger than previous versions of signaling due to *store* and *send* address code in a bloc system.

Signaling System No. 6

The signaling system No. 6 is also a standard developed by the CCITT (which is ITU-T at present). This is also known as CCITT6 or CC6 signaling system. The code capacity of this signaling system is much higher than the previous ones. It is highly suitable for transit and terminal signaling. The standard was adopted in 1968, and the testing of the system began in 1970–1972. It was also tested for the local telephony system other than the international signaling system. One of its variants was also used in the local telephony signaling system in the USA.[63]

This signaling system is standardized under Q.251–Q.300 recommendations from ITU-T. It is a type of very preliminary version of the CCS system but cannot be recognized as the full-fledged CCS system as its successor SS7 is known for. It is extensively used for international analog telephone systems and international satellite links.

The main features of this signaling system are as follows:

- It uses a common link for signaling.
- It is used for both-way signaling.
- It supports two modes of operations – quasi-associated, and associated.
- It uses a signaling unit for sending out signals.
- Each signal bit consists of 28 bits, which includes 8 check bits.
- It is suitable for both analog and digital systems.
- For analog systems, the speed of this signaling system is 2,400 bps.
- For digital systems, the speed of the signaling is 4 kbps.
- It supports error check and signaling re-send option.
- Address is sent out in overlap and bloc modes.
- It supports much higher speed than its predecessors.
- It has a huge capacity of signal codes as compared to previous signaling systems.
- Supports eco suppression in the links.

- It is suitable for all types of telephone circuits such as speed interpolation circuits and others.
- CC6 signaling is also suitable for satellite links.

COMMON CHANNEL SIGNALING (CCS)

CCS is a category of signaling system in which the signaling protocol uses a separate channel for sending the control signals. In other words, the data and voice signals over the transmission channel use the control functions transmitted over a separate channel (not on the same channel on which the data and voice travel). The modern signaling systems (used for the public switched telephone network (PSTN)) and cellular mobile networks use the CCS system like the SS7 signaling system.

The use of CCS in analog systems is done through signaling system No. 7. The SS7 is also extensively used for digital systems. Signaling system No. 7 will be extensively elaborated in the next chapter which is dedicated to digital signaling systems. Now, we will know about the signaling system No. 7 used for analog telephony systems in the past.

Signaling System No. 7

SS7 is a very powerful signaling system that supports analog signaling, digital signaling, and advanced digital signaling based on intelligent networks (INs). The architecture of the signaling system 7 is very flexible and supports network adaptability. The architectural details of this signaling system will be discussed in the next chapter.[64]

The SS7 signaling protocol is also known as C7, CCITT7, and CCS7 system in the common terminologies in the field of telecommunication. This is an ITU-T standard defined by the CCITT committee. The recommendation for the definition of specifications of this signaling system includes the Q.700 series. It was first standardized in 1975, and later on, it was implemented as an analog PSTN signaling system across the world. It became the first signaling system extensively adopted across the world for all types of telephony signaling systems, i.e., local, national, and international telephony systems.

The salient features of the SS7 protocol are listed in the following:

- It supports Plain Old Telephone Service (POTS) and PSTN systems.
- It is capable of supporting both analog and digital telephony.
- It uses a separate channel for signaling and voice or data in digital systems.
- It is a comprehensive telephone protocol that is capable of establishing, supervising, and tearing down the telephone call over a hierarchical telephone system.
- It supports associated and quasi-associated modes of signaling.
- Analog SS7 system layers include MTP1 (Message Transfer Part Level 1), MTP2, MTP3, and TUP (Telephone User Part). (The details of these layers will be discussed in the next chapter.)

ANALOG TELEPHONE SWITCHING PROTOCOLS

Telephone switching is the process of connecting an incoming telephone call to the outgoing trunk or telephone line. The incoming and outgoing telephone lines are physical or logical telephone calls to be connected to the end-users of the telephone service. Let us know a little bit about the need for switching in the telephone systems, before talking about the telephone switching techniques and protocols used in analog and digital telephony systems.

Flash back to 1876, let us take into consideration the first telephone system that was created and patented by Alexander Graham Bell. This telephone system consisted of two telephone devices between two end-users directly connected through a physical telephone cable. This system consisting of just two users directly connected to each other did not require any kind of switching technique and its respective protocol at all.

Now, let us consider a telephone system consisting of six telephone users located at two different towns and they want to talk to each other on the telephone system. What would you need to connect them? You need to lay down a dedicated telephone line between each telephone located in different towns as well as within the same town. You need 15 telephone lines to connect between those 6 telephone users. Then, what about a telephone network of four cities with 500 users in each city to be connected with each other? You would need (2,000×1,999)/2=1,999,000 dedicated lines to connect those telephone users located in four different towns. The number of lines needed to connect such a telephone system of "N" telephones would be $N*(N-1)/2$. This is in fact, similar to a fully meshed network setting. Hence, this is not a commercially viable solution at all (as, quite impractical when the number of users increases).

Hence, the new technique of switching system or switching exchange emerged just a couple of years after the invention of the telephone system. Till now, numerous types of manual switching, semi-automatic, and automatic switching systems have been introduced in telephone systems. The major types of switching protocols used in analog telephone systems are listed in the following:[65]

- Manual Switching System
- Automatic Telephone Selector System
- Step-by-Step Switching System
- Crossbar Switching System
- Semi-Electronic Switching System

Manual Switching System

In 1878, the first manual switching system was used in which an operator would use patch cords to connect two different telephone lines on a telephone board. The manual switching increased the efficiency and capacity of the telephone networks significantly. The lines between two cities or small towns to connect multiple numbers of telephone users were possible with the help of manual switching.

In the manual switching system, a board with a grid of the telephone numbers is used in a small exchange room. The exchange is controlled and operated manually by a telephone

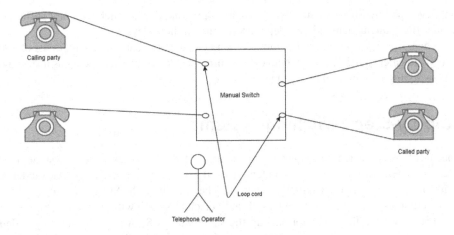

FIGURE 2.9   Schematic diagram of the manual switching system.

operator, who would use a patch cord to connect the incoming call to the outgoing number located in his or her exchange. Then when the conversation between two users is completed, the light bulb would turn on to show the line is free. At the end of the conversation, the telephone operator would log the billing manually by calculating the time of the call.

In this process, the efficiency of the trunk lines between two long-distant cities and towns increased significantly. A few lines would suffice in catering to the needs of a large number of users at both ends of the network.

The schematic diagram of the manual switching system is shown in Figure 2.9.

The incoming call from any user would be sensed through light or buzz, and the operator would talk to the caller for his request for connection to the desired telephone number. The operator would connect that particular number on the board through a telephone patch cord connecting those lines. This was the first manual technique used for analog telephone systems in the world.

Automatic Telephone Selector System

Automatic telephone selector system was a new technique for switching through an electromechanical selector device. This system was introduced and patented by Almon Brown Strowger in 1889, just 11 years after the manual system for switching was launched in 1878. In this type of analog telephone switching system, the 10 arches of 10 sets of numbers were stacked to form an outlet of 100 numbers. The inlet of the switch would move vertically and horizontally to choose the desired line located on the 10 stacked arches of 10 numbers each.[66]

The inlet system of the Strowger system would move vertically to choose the position of the arches and then would move horizontally to choose the position on the arch. Thus, the desired number on the outlet of the system would be connected to the incoming call from the inlet device. This system is also referred to as a two-motion selector system, in which the motion of the selector would be controlled by the inlet dialed digits from the dialer. The first digit dialed by the subscriber would control the vertical movement of the selector, while the second digit would control the horizontal motion of the selector. This tool developed by Strowger supported just 100 outlet numbers, which were not sufficient to cater the growing demand for telephone systems in those days.

Then, this system was further modified and enhanced to support 10 multiples of 100 systems to cater to the 1000 numbers of switching nodes, which was known as a step-by-step switching system as elaborated next.

Step-by-Step Switching System

The step-by-step switching system is an extension of the Strowger selector system. This selector system was invented and patented by three American scientists: Alexander E Keith, John Erickson, and Charles J Erickson. They modified the Strowger switch, which consisted of a single selector known as "Final Selector" switch with two selectors.[67]

They used the final selector exactly the same way as Strowger did, but they added another selector known as group selector, which would use 10 different selectors to cater more switching needs. The addition of a group selector would support the switching of 1,000 subscribers. This system was much more flexible in its assembly, and it was possible to scale up the capacity by adding more group selectors. But, this would increase the bulkiness of the mechanical gears and motion in the system. At that particular time, this system was capable enough to support the growing demand for telephone lines in the telephone exchanges across the world.

The first pulse dialed by any subscriber for connecting to any desired number would be used for selecting the right group selector. The dialed numbers would raise the selector by one step from digit 1 to 9 and then 0 as the tenth step up. The next digit dialed would control the vertical motion of the "Final Selector", and the third pulse would control the horizontal motion of the "Final Selector" for switching the line to the right number of on outlet of the switch.

It is very important to note that the motions of the group selector and final selector were replaced by some electrical devices such as registers and translators, which were based on the wired logic connections, later on. The dialed digits would be stored in those registers, which would control the motion of the selectors. The translators would understand the digits stored in the register to control the motion of selectors through registers.

Crossbar Switching System

Crossbar switching system is another electromechanical system. This switching system also uses the concept of selectors but with a different physical structure of the switching system. This switching system was invented and patented by G.A. Betulander and Nils Palmgren in 1919. Both of those two scientists belonged to Sweden. They also founded a small company after their invention. They were working on this project since 1910 for the improvement in the switching system. They applied for the patent in 1912, which was later approved in their favor.[68]

The manual switching was one step forward to the efficient use of the trunks laid down between two cities and towns located at long distances, but it was fully dependent on the manual operations, controlled and managed by a human operator in each exchange. This was a very slow process and would need a huge manpower fully trained for the operations and management of the telephone exchanges (that one would operate manually).

Crossbar switching became one of the biggest hits in the telephony systems, and it continued into operations for many years to come. The popularity of this switching system reached a peak point during the periods of the 1950s and 1980s. After the 1980s, digital telephone switching started taking popularity and started replacing analog telephone switching systems systematically. The first crossbar switching system was installed in 1938.

The assembly of a crossbar switch is based on two systems of bars placed in cross, as the name of the system suggests. The structure consisted of horizontal bars is referred to as "selecting bars" and the vertical bars, which are known as "hold bars". The motion of the bars was controlled by the electromagnets connected with those bars. The electromagnets would receive the signals from the control devices to get energized accordingly and would move the bars in their respective directions.

The controllers of the electromagnets in this system are known as "markers". The direction of movement of selecting bars is either up or down. The *hold bars* rotate for making contact with the *selecting bars* in the system. This type of switching would require huge power for operating the switches and the maintenance of those switches was also huge as compared to the advanced switches that came into existence later on.

The image of a crossbar switch used in the mechanical switching of the telephone system is shown in Figure 2.10.

After the deployment of this system in numerous telephone systems, the switching system continued to operate in many countries across the world, especially in the European countries during the entire span of the 20th century. The copyrights of this switch were taken over by the Ericsson Corporation by acquiring the company founded by the inventors of this switch.

FIGURE 2.10 Crossbar switch (Flickr).

Semi-Electronic Switching System

We can say that the semi-electronic switching system was the preliminary switching system based on modern electronics. This system would use flip-flops, gates, and other electronic materials in designing the logic circuits of telephony switching. The reed relays were used in this initial version of the telephone switching system for transferring the analog signals coming from the input telephone device to the output devices.

The structure of a reed relay was based on a glass envelope. That glass acting as a tube would house contacts made from ferrous metallic reeds. Those ferrous reeds would be placed into the glass envelope hermetically isolating environment inside the glass. Then, this envelope of housing metallic reeds would be placed inside the electromagnet generating coil. When the current passes through the coil, the ferrous metallic contacts would close and input and output pair of connections would be connected.

Later on, this semi-electronic system was replaced by analog gates and then transistors and other modern electronics. With the installation and commissioning of the first digital switching switch in 1976, the new era of digital telephone switching systems started. Different versions and types of digital switching techniques were introduced in the telephone system.

CIRCUIT SWITCHING

Circuit switching is one of the most popular and extensively used switching systems in the telephony system. Circuit switching is a mechanism of transmitting the incoming telephone line to the outgoing line through either a physical circuit or a logical circuit. The use of a physical circuit was the core feature of this switching system in analog telephony.

All of the aforementioned switching systems would use the circuit switching mechanism. The circuit switching was also capable of handling the digital signals and would use the physical circuits more efficiently. The details of circuit switching will be mentioned in the next chapter.

Sample Questions and Answers for Chapter 2

Q1. What are the major codes used in optical telegraph systems?

A1. The major codes used in optical telegraph systems are:
- Chappe code
- Wigwag code
- Edelcrantz code

Q2. What is Morse code? Explain.

A2. Morse code is a method used in telecommunication to encode text characters as standardized sequences of two different signal durations, called dots and dashes. It is named after Samuel Morse, one of the inventors of the telegraph. Different letters, numbers, and punctuation marks were coded with different sizes of dashes, spaces, and dots. The variable sizes of dashes were generated by sending the electrical current for variable times at the point of sending the signals, which would produce voltage at the receiving end through armature (in electrical engineering, an armature is the component of an electric machine that carries alternating current) and the coil would drag the printing head on the paper to mark spaces, dots, and dashes of different patterns. The definition of those spaces, dots, and dashes was defined under the Morse code of electrical telegraphy. The Morse code encoded the numbers from 0 to 9 and alphabetic letters from A to Z. The total number of initial communication symbols and letters that were encoded through the Morse code was 36. Later on, different signs and special characters were defined in different versions of the Morse codes.

Q3. What are the purposes of telephone communication protocols?

A3. The purposes of telephone communication protocols are listed as follows:
- Alerting for incoming call
- Addressing destination
- Supervising idle telephone lines
- Informing the status through response

Q4. Name the two major categories of interoffice signaling protocols.

A4. The interoffice signaling protocols can be divided into two major categories in terms of the status of the channels used in the signaling:
- Channel-Associated Signaling (CAS)
- Common Channel Signaling (CCS)

Q5. Why is it impractical to have a direct and dedicated line between each pair of telephones in a real-life telephone system?

A5. The number of lines needed to connect such a telephone system of "N" telephones would be $N*(N-1)/2$. This is in fact, similar to a fully meshed network setting. As the number of N increases, it would be practically impossible to make those connections. For instance, a telephone network of four cities with 500 users in each city to be connected with each other would need (2,000×1,999)/2 = 1,999,000 dedicated lines (to connect those telephone users). Hence, this is not a commercially viable solution at all (as, quite impractical when the number of users increases).

Introduction to Digital Communication Protocols

3

The digital communication began with the advent of modern electronic devices such as gates, transistors, flip-flops, registers, and others in the late 1970s. The digital communication started with digitized telephony and digital fax machines that were extensively used in the 1980s and onward. The change in digital technology was so rapid that the fax machine started getting replaced by emails with the advent of modern computer systems and the Internet. The telephone systems have also passed through very rapid changes in technologies during the past few decades.

All those changes in the digital transmission of telephone and data are governed by the modern digital communication protocols, which will be discussed in this chapter.

DOI: 10.1201/9781003300908-4

DIGITAL TELEPHONE SWITCHING
AND ITS TECHNIQUES

Digital telephone switching is a type of switching of the incoming call to the outgoing call through electronic gates and data storage registers. The switching matrix and the logical gates establish the connection between the two subscriber lines located on the two sides of the switch. The digital switching uses the PCM word, which consists of 8 bits. The information in the PCM word is collected and re-arranged to form a complete set of information required for locating the destination address. Once the destination address is known, a PCM slot is allocated for the subscribers for sending out information in different intervals. The PCM uses an 8 kHz sampling frequency for coding the information requested by the dialer subscriber and sends it out to the dialed destination.[69]

Digital telephone switching can be divided into three major techniques as listed in the following:

- Space division switching
- Time division switching
- Time–space division switching (hybrid)

Space Division Switching

Space division switching is the type of switching in which the input information is switched and transmitted to the output subscriber at the same time when the input data are sent. A dedicated circuit for a certain period is established for sending and receiving information simultaneously. In other words, the incoming slot of the PCM is connected to any available outgoing PCM slot. Separating them spatially not in terms of time is known as space division switching.

In the beginning, this switching was designed for analog switching but later on, it was used for the digital switching system too. In this type of switching, a periodic slot in terms of time is not used, but rather a direct and dedicated slot is connected to the destination without any storing of the incoming information and waiting for the time slot cycle to come. This is also referred to as cross-point switching, which is faster or instant.

Time Division Switching

In time division switching technique, a certain time slot of a PCM-modulated trunk is dedicated for the outgoing line or connection. The incoming information is stored in the register, and it waits for the time when the cycle of the slot reaches to transmit the stored information. The slots are distributed in terms of time domain. A PCM signal from the incoming side is stored in the register and an outgoing free slot is used for cyclic transmission of the incoming information.

As compared to space division switching, time division switching is purely used for digital switching by allocating the dedicated outgoing circuit on the basis of a time slot, which will be used for a particular subscriber during the entire call duration. In this switching technique, the cross-points are shared for a short period of time periodically, which is not done in the case of the spatial or space division switching technique.

Time–Space Division (Hybrid) Switching

To improve the efficiency and speed of the trunks in the telephone systems, the combination of space and time division switching techniques is used to form a hybrid type of technique for telephone switching. In this technique, the time division switching is used for input and output switching, while the space division switching is used in the between the input and output switching techniques to reduce the number of cross-points.[70]

Circuit Switching

Circuit switching is the general name of dedicated line switching in which a circuit is dedicatedly allocated to the call from origination to destination devices through different techniques of switching such as time division or space division switching. Circuit switching is also known as connection-oriented switching. This dedicated circuit cannot be used by any other subscriber despite the fact that the data on the link may be very low or no data is traveling at a certain period of time. This entire end-to-end process of circuit switching is divided into three major processes as listed in the following:

- **Establishing connection** – In this process, the dialing subscriber dials the destination number and switch allocates the circuit from originating point to the destination point passing through one or multiple telephone exchanges.
- **Transferring data** – The second process of circuit switching process is transferring data on a dedicatedly assigned circuit across single or multiple telephone exchanges to the destination subscriber.
- **Disconnecting circuit** – The last step of the circuit switching process is disconnection of the dedicated connection between two subscribers. This is done when any one of the subscribers tears down the circuit.

Circuit switching has become a very slow and resource-wasting technique compared to the latest modern technologies, which are relatively much faster and more efficient in transmitting, receiving, and processing the data. The circuit switching remained as the core data and voice switching technique for many decades before the increase of power and efficiency of packet switching technique in the modern networks or the Internet and data communication.[71]

The schematic diagram of a circuit-switched connection between two subscribers is shown in Figure 3.1.

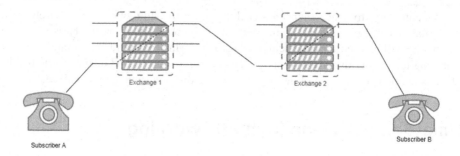

FIGURE 3.1   Schematic diagram of circuit switching.

The main disadvantages of circuit switching over the latest IP-based packet switching system are listed in the following:

- Dedicated connection is not used efficiently; so, it results in a wastage of valuable transmission resources.
- It takes relatively longer for the establishment of a connection.

Packet Switching

Packet switching is one of the latest types of switching techniques, which uses small packets of data for transferring data from one point to another one. In packet switching, the information or data of any message originating from the destination and destined for a certain node is divided into multiple small units of data, which are known as the packets of data. The data packets are switched to the destination through multiple paths and rearranged at the destination to constitute the original data of the message. Packet switching is also referred to as connection-less switching in which no dedicated circuit or route is used for the data transmission.

The modern Internet is heavily based on the packet switching network. The modern telephony system such as Voice over IP (VoIP) also uses packet switching at the backend transmission. The packet used for packet switching consists of two major parts:[72]

- The headers of a packet
- The payload of a packet

The header of a packet consists of destination address, source address, sequence number, and other information, which are used by the routers and other nodes of the network to pass/forward the packet to its destination. The payload consists of a part of data of the original message and headers of the lower layer protocols such as physical and data link layers.

A message or information, which is required to be transmitted to the destination, is divided into multiple packets of data. Each data packet is routed through different routes consisting of multiple or even single network node in between the source and destination.

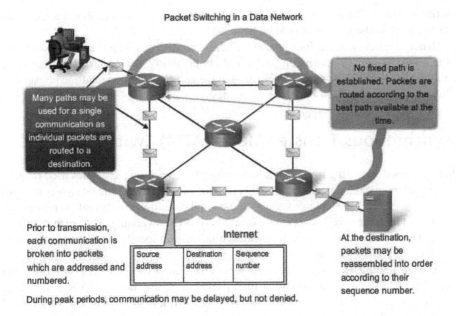

Packet Switching in a Data Network

Many paths may be used for a single communication as individual packets are routed to a destination.

No fixed path is established. Packets are routed according to the best path available at the time.

Prior to transmission, each communication is broken into packets which are addressed and numbered.

Internet

Source address	Destination address	Sequence number

At the destination, packets may be reassembled into order according to their sequence number.

During peak periods, communication may be delayed, but not denied.

FIGURE 3.2   Packet switching network (Flickr).

The packets reach at destination in a random order because they travel from different routes. It is the responsibility of the upper-level protocol to reorganize the packet in a sequence and constitute the original information at the destination.

A comprehensive diagram for packet switching from the point of origination to the destination is shown in Figure 3.2.

In this figure, you can see that the packet traveling from source to destination takes different routes or paths. There is no dedicated connection for transporting the entire message in sequential order as the circuit switching does.

Message Switching

Message switching is another form of network switching used for data transmission. In the message switching technique, the entire message originating from a source is collected and stored in the memory storage. The entire message is then forwarded to either the destination node or the nearby node, which can help forward that to the destination. The next node also collects the entire message and stores it in the buffer and waits for the right time, when the resources are available for transmitting the entire message to the destination.[73]

This type of switching was a good alternative to circuit switching to save transmission resources and improve the network performance substantially. But, the major downside of this switching technique was that every node in the network would require a huge storage space to save the messages before delivering them to the destination or to the nearby node to the destination. Another disadvantage of this switching system was the slowness of the data transmission. The message would be stuck for relatively long periods in certain cases when the load on the network would be huge.

Hence, the best way out for efficient switching in modern computer-based communication networks is packet switching, which offers flexibility, performance, and easiness. The message switching was extensively used in the text messages on landline phones and cellular mobile phone networks in the recent past.

Asynchronous Transfer Mode (ATM) Switching

IP-based packet switching is one of the most powerful switching techniques used in modern data communication networks. But, this form of switching is a little bit slower in terms of data throughput. When a huge volume of data in the multi-giga-bits scale is required to be transmitted in the core network, you would need ATM switching. The performance of IP would not be a favorable choice for the data transmission speed as in that case, the ATM switching comes as superior.

ATM switching uses cells of constant length for transporting the data from one point to another. ATM is a faster switching than packet switching due to efficient switching fabric and virtual identifiers used for transporting the information. The cell of an ATM switching includes:

- Virtual Path Identifier (VPI)
- Virtual Channel Identifier (VCI)

ATM switching uses space switching in the switching fabric, which transfers the data much faster and in bulk volume at the switching fabric based on VPI and VCI identifiers.[74] It uses multiple buffers for queuing the data at input, output, and center. The major queuing disciplines used by ATM switching for faster transmission include:

- Input queuing
- Output queuing
- Central queuing

These queuing help the ATM switching to collect, forward, and process multiple cells for creating better efficiency of the switching network.

BASIC DATA COMMUNICATION PROTOCOLS

Data communication is the domain of communication in which different technologies related to computing and communication are used for the transmission of any kind of data from one place to another by using different elements of data communication systems. Those elements of data communication systems can be both software elements

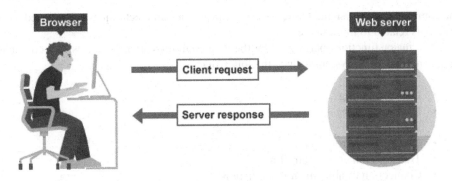

FIGURE 3.3   Pictorial presentation of a simple data communication network (Flickr).

and hardware elements. The basic communication system consists of the following elements:

- Sender
- Receiver
- Message
- Medium
- Protocols

The communication protocols and message are software-based elements, and the rest of the components are hardware-based. Medium can be of different forms and formats including the cables, optics, air, and others.[75]

Figure 3.3 shows a simple presentation of a data communication network. In this figure, the browser on the computer used by the end-user is called the receiver, which receives the information through the medium. Here in this picture, the medium is not clearly depicted; it can be one or more types of mediums used from the source or sender of the message to the receiver. The mediums can be copper cables, fiber cables, air interface, and others. The request and responses shown inside the arrows show the activities performed by the protocols of data communication. Many types of protocols are involved in the entire communication of the data from the receiver to the sender.

We discussed numerous forms of telephone and telegraph protocols, which are also communication protocols one way or the other. However, in the modern terminologies, those protocols are specifically referred to after their respective perspective of the fields such as telephone protocols, telegraph protocols, etc.

It is very important to note that data communication involves numerous types of digital and analog converters at transmitting and receiving nodes. Similarly, a large number of processes such as multiplier, amplifier, booster, regenerator, and many other processes related to security, compression, encryption, and other factors are also involved in this entire process of data communication. Different protocols come into play at different processes to carry out the desired activity required for a certain functionality of the communication system. Those processes, protocols, and techniques are referred to as protocols related to those processes.

The major functions performed by the data communication protocols in a complete data communication system include the following:

- Data formatting
- Data sequencing
- Data routing
- Data flow controlling
- Transmission error controlling
- Connection establishment/termination
- Data security and integrity
- Transmission order or precedence
- Log of events (information)

Hence, any technique or protocol dealing with the aforementioned functionalities of a protocol in the data communication system is considered as a protocol to deal with that particular domain of data transmission systems. Let us talk about the major processes, techniques, and protocols used in the data communication with more details.

Analog to Digital Conversion

Analog to digital conversion commonly referred to as ADC is a process of converting the analog input of signals generated by voice, video, light, electric, temperature, and other sources into digital signals.

Analog signals are variable in amplitude, frequency, and other parameters, which are more difficult to handle and transmit for longer distances. Those signals are very prone to environmental factors and other similar kinds of signals traversing in the adjacency. To achieve the most powerful characteristics of digital signals, the analog signals are converted into digital signals through the ADC process in the field of digital signal processing (DSP) in electronic engineering.

The ADC is done through the combination of different electronic components such as diodes, transistors, gates, flip-flops, registers, and others integrated into an architecture commonly known as electronic architecture. This architecture of ADC is known as an integrated circuit (IC). There are a wide range of architectures used in ADC based on different performance, accuracy, efficiency, cost, quality, and many other factors.

The most commonly used architectures for ADC are listed in the following:[76]

- Successive approximation register (SAR)
- Flash/paralleled architecture
- Mesh architecture
- Sigma–delta (SD) architecture
- Pipelined architecture
- Delta-Encoding (DE) architecture
- Intermediate FM stage architecture

FIGURE 3.4 MPC3208 ADC chip with eight input ports made by Microchip (Flickr).

The digital to analog conversion (DAC) is done in two major stages. Those two basic steps are also electronic digital processing processes, as mentioned in the following:

- Signal sampling process
- Signal quantization process

An electronic circuitry for ADC manufactured by Microchip Inc. is shown in Figure 3.4.

In the process of analog to digital signal conversion, first of all, the analog signal in many different forms such as light, voice, temperature, pressure, and many others is input through suitable analog sensor ports or interfaces into the architecture. In the first step, the signals are sampled at a certain rate. The sampling rate is determined by certain laws. The most commonly used law for analog signal sampling is known as Nyquist–Shannon law. The sampling rate is directly proportional to the quality of the digital signal. The higher the sampling rate, the better the quality of digital signals.

The second process of ADC is the quantization of the signals. In this step, the digital samples are converted into the patterns of binary digit systems. It can be used in different types of systems such as 3-digit, 7-digit, 15-digits, and so on. The quantization level and throughput of the system are affected inversely with the changes in any of those two components of the ADC process.

It is very important to note that in most of the time in data communication systems, the ADC and DAC are performed with a certain algorithm. This helps consolidation of the communication system through seamless transmission from both ends of the communication systems. Nowadays, multiple sampling schemes are extensively used for the sake of better quality and throughput. The most popular sampling schemes used in different applications of ADC include:

- Over-sampling
- Under-sampling

The over-sampling is done for increasing the quality and accuracy of translating the analog signals into digital signals. Similarly, under-sampling is also becoming a common practice for different applications.

Digital to Analog Conversion

The DAC is an electronic signal processing method in which the digital signals in the form of streams of zeros (0s) and ones (1s) coming from an electronic device are converted into the variable and irregular types of analog signals. This process is exactly the opposite of the ADC.

DAC is extensively used in different fields of technologies such as telecommunication, audio, video, mechanical, and other applications. In telecommunication systems, high-speed DAC and ultrahigh-speed DAC are used. High-speed DAC is used in mobile communication and ultrahigh-speed DAC is used in optical communication systems, which transmit huge data in bulk.[77]

Figure 3.5 shows the chip of single-channel 12-bit DAC converter manufactured by Microchip Inc.

There are many types or architectures of DAC systems. A few very important ones are:

- Cyclic or successive-approximation DAC
- Pulsewidth modulator DAC
- Interpolating delta–sigma DAC
- Thermometer-coded DAC

FIGURE 3.5   Microchip's MCP4921 DAC chip (Flickr).

- Hybrid DAC
- High-speed DACs
- Ultrahigh-speed DAC

Highway Addressable Remote Transducer (HART) Protocol

Highway Addressable Remote Transducer (HART) is a communication protocol that is used for transmitting digital signals over analog wiring systems among the Internet of Things (IoT) devices and the control systems. HART protocol was released at the beginning of the 1980s. This protocol was consistently getting grounds in the field of remote monitored devices. The main characteristics of the HART protocol are listed in the following:[78]

- Initially, it was launched as a proprietary protocol.
- This protocol was made open source in 1986.
- It supports traditional equipment system 4–20 mA analog instrumentation.
- It supports three types of HART commands as mentioned in the following:
  - Common practice commands
  - Universal commands
  - Device-specific commands.
- It is managed by the independent community known as HART Communication Foundation.
- It supports two modes of operations, as listed in the following:
  - Multi-drop mode
  - Point-to-point mode.
- A large number of companies manufacture HART-enabled devices for modern IoT ecosystem across the globe.
- The communication packet of a HART protocol consists of the following fields and length of the fields:
  - **Preamble field** – This is a 5-to-20-byte field used for the synchronization of communication.
  - **Start byte** – It is used for specifying the master number. It is a one-byte field.
  - **Address field** – This field consists of 1–5 bytes, which specify master, slave, and burst mode features.
  - **Expansion** – It is a 0–3 bytes field, which is defined in the start byte.
  - **Command** – This is a single byte field, which specifies the command value.
  - **Data field** – This field can vary between 0 and 255 bytes. This is the core payload field of the communication, which defines command parameters and other characteristics of the command to be issued.
  - **Checksum** – It is a single byte field to check for the errors in the communication.

The HART communication protocol is becoming an important component in the modern IoT environment powered by the modern control systems dealing with analog as well as digital signals powered by the modern Internet and wireless technologies.

Modem Data Communication

Modem data communication dates back to the early 1950s, when the American military used the Semi-Automatic Ground Environment (SAGE) system for data communication purposes among different airbases, satellite stations, and other military installations. This communication system remained in use for a very long period. This system is used in some parts of the world even today. This technology powered by different techniques and protocols has evolved over seven decades so far.

The modem data communication started with the serial communication of bits at very low baud rates starting from just a few bits to thousands of bits as the technology and protocols advanced. The major algorithms and techniques used in modem communication include:

- ASK
- FSK
- PSK
- QAM
- Combination of multiple shift keying schemes

The purpose of using different types of techniques is to increase the efficiency of DAC and ADC conversions, which are the fundamental processes used in modem data communication. By using any kind of modulating scheme, the number of bits coded in a data symbol transmitted over the serial port is increased. This is the main reason for using efficient techniques of modulation. Initially, one single symbol would be coded for just one bit of information; thus, a very limited speed of data communication would be realized in such communication. The use of multiple schemes of coding together has increased the modulation efficiency significantly in the modem communication systems. The advent of modern modulating schemes like frequency-division multiple access, OFDM (orthogonal FDM), and others has increased the speed of modem data communication significantly.

The term MODEM is derived from two words: modulation and demodulation. The purpose of modem is to modulate the analog signals into digital signals and digital signals into analog signals. The communication within a computer is normally done digitally. Similarly, the communication within the existing telephone exchanges has become fully digital but the last-mile network is still analog through local loops. To use the analog local telephone lines and digital systems, a modem is used to convert the digital signals originating from a computer to analog ones. Likewise, it is required to convert the analog signals received from the local loop telephone lines into digital signals.

This same process also takes place at the local telephone exchange where the analog signal received from the telephone local loop is converted into the digital signals, and the digital signals emerging from the digital exchange and servers are converted into the

FIGURE 3.6   Old modem and telephone (Flickr).

analog signals to be transmitted over the analog loops of the last-mile network. A photo of modem and telephone used in the past is shown in Figure 3.6.

There are numerous protocols that were in use since the beginning of modem-based data communication systems. A few of those major protocols used for modem data communication are listed in the following:[79]

- **Bell 101** – used FSK technique with a baud rate of 0.1 kilobits per second.
- **V.21** – This is the first protocol approved by ITU in the V series. It used FSK modulation and was able to transmit 0.3 kilobits per second data.
- **V.22** – It is an advanced version of V.21 with a data rate of 1.2 kbps and used QPSK modulation.
- **V.22bis** – It is an advanced version of the previous standard recommended by ITU. The data transmission capacity was 2.4 kbps and used QAM modulation.
- **V.32** – This protocol developed by ITU specifies the interfaces and modulation with a capacity of 9.6 kbps data rate.
- **V.32bis** – This protocol uses Trellis modulation, and the data bit rate is 14.4 kbps.
- **V.34** – This protocol was capable of transmitting data at 28.8 kbps and used trellis modulation. Later on, this modem was enhanced for 33.6 kbps with 3,429 baud data throughputs.
- **V.90** – This is the first protocol to handle data transmission rate up to 56 kbps with downlink speed and 33.6 kbps uplink speed.
- **V.92** – This protocol was able to enhance the uplink speed up to 48 kbps and downlink speed 56 kbps with 8,000/8,000 baud rates.

As we know, modem communication uses normal telephone lines for communication purposes. Those telephone lines have four cables used for signals and power for ringing, tones, and device energization. A telephone uses four-pin connectors known as RJ11, which is referred to as Registered Jack No. 11 (RJ11). An RJ 11 connector is directly connected to the modem installed on the computer or sitting on your computer. The modem is connected to the computer through other physical connections such as serial ports.

SERIAL COMMUNICATION PROTOCOLS

Serial communication is a very vital and one of the oldest forms of communication protocols. They are used for the communication between different peripherals of computers such as printers, fax, camera, keyboard, and others as well as among different analog nodes integrated into an automated control environment in any industrial unit. Serial communication protocols can be divided into two categories as listed in the following:[80]

*   Asynchronous serial communication protocols
*   Synchronous serial communication protocols

Every category has different types of protocols used in the industries, especially in the communication industry. A few very important asynchronous and synchronous serial communication protocols are mentioned with more details in the following sub-sections.

RS-232 Protocol

Recommended Standard 232 precisely referred to as RS-232 is a serial communication protocol. It is extensively used for connecting different types of computer peripherals such as monitors, printers, and other machines connected for process automation. Two physical interfaces known as DB-9 and DB-25 support RS-232 protocol. The detailed diagram of the DB-25 connector is shown in Figure 3.7.

The main characteristics and features of the RS-232 communication protocol are listed in the following:

*   It is a point-to-point communication protocol.
*   It supports one device at a time.
*   It supports two-way communication with line control commands.
*   The supported distance is up to 15 meters.
*   The supported data rate is up to 96 kbps.
*   Transmission is digital in the form of zeros and ones.
*   Zeros and ones are characterized by different levels of voltages.
*   Zero is known as space, which is transmitted through +3 to +15 voltage.
*   One is known as mark, which is transmitted through −3 to −15 voltage.
*   The connector of RS-232 protocol is known as the DB-9 connector.
*   Connector is available in male and female assemblies.

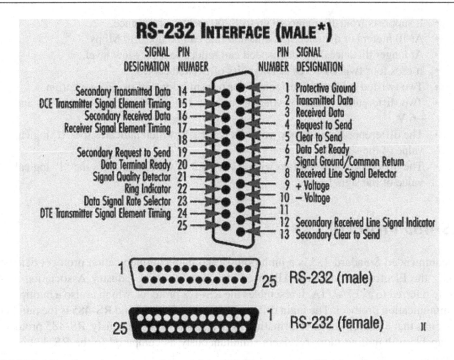

FIGURE 3.7 Connector pin positions of the DB-25 connector (Flickr).

- Another important connector that supports RS-232 protocol is the DB-25 connector.
- This protocol is extensively being replaced by the USB protocol, which will be discussed in the next topics.
- This protocol is also used in CNC machines, packet loss concealment (PLC) control systems, and other industrial applications in many industries.
- It is defined under the specification of EIA/TIA-232.
- In telecom systems, it is used for the communication between data terminal equipment (DTE) and data circuit-terminating equipment (DCE).

RS-422 Protocol

Recommendation Standard 422 is precisely referred to as RS-422 serial data communication protocol. The working principle of this protocol is almost similar to the RS-232 protocol. It is capable of transmitting signals to up to 10 devices simultaneously. Another difference between RS232 and RS422 is that the latter uses differential signals. This protocol can support the DB-9 connector as well as terminal blocks. The main features of RS-422 are listed in the following:

- Only device transmits at a time and up to 10 devices can receive signals.
- It uses differential signals for transmission.

- It supports from 10 meters to up to 1,200 meters of distance.
- At 10 meters of distance transmission rate can be up to 10 Mbps
- At longer distances, the data speed can reduce to a very low level.
- It uses four twisted wires or two pairs of twisted cables.
- Two twisted cables are designated for transmission and two for reception.
- Two differential points of voltages used in this physical protocol are +6 and −6 V.
- The difference between two voltages is +0.2 V, which indicates the "0" logical value of the signal.
- The difference between two voltages is −0.2 V, which indicates the "1" logical value of the signal.

RS-485 Protocol

Recommended Standard 485 is a multipoint serial data communication protocol defined under the Electronic Industries Alliance/Telecommunication Industry Association precisely referred to as EIA/TIA. It resembles the RS-422 protocol, which is also a multipoint communication protocol. The main difference between RS-422 and RS-485 is the number of points that can receive the information from a sender simultaneously. RS-485 protocol is capable of handling more receivers simultaneously as compared to the RS-422 communication protocol.[81]

The main features and characteristics of the RS-485 serial communication protocol are mentioned in the following list:

- It is a long-distance serial communication protocol ranging up to 4,000 feet.
- The data rate at 4,000 feet distance decreases to about 100 kbps.
- The data rate at 50 feet distance is about 10 Mbps.
- It is defined under EIA/TIA-485 specifications.
- It supports two wire connections on a bus to connect multiple devices simultaneously in half-duplex communication mode.
- For full-duplex communication, a four-wire network is used.
- It is a multi-drop communication protocol, which supports up to 32 receivers and transmitters in a communication data network.
- It is extensively used in the centralized control systems of the network connecting numerous devices and equipment in home, office, or industry environment.
- It is able to control the data collision in the network without any external software support.
- It is also able to find the bus fault conditions in the communication network.
- It supports master/slave node configuration supporting the feature of going into tri-state (power-off) mode when not transmitting any data.
- Some of the modern RS-485 compliant hardware changes the power-off condition automatically after sending a command or requested data on the network.
- This protocol supports half-duplex mode of communication in two-wire mode and full-duplex of communication mode in the four-wire configuration.
- It uses the balanced signaling.

- It is a robust and highly popular protocol in data communication-based device controlling systems that require high data rate at short distances and low-data rate a remote long distance in hybrid types of data communication networks.
- It is one of the highly reliable data communication networks used in the communication of networks consisting of different types of devices simultaneously.
- In certain cases, this protocol can also be configured in a four-wire format.
- Ground wire is necessary for both configurations, i.e., four-wire and two-wire configurations.
- RS-485 also supports repeaters for a long-distance network.
- The conversion of RS-485 into RS-232 protocols is also possible through RS-485/RS-232 converters available in the market.
- Extensive usage of this protocol is found in industrial automation, computer automation, commercial aircrafts cabins, PLCs, theaters, studios, railways, and many others.
- The signal format of the RS-485 interface is determined by two signals known as A and B along with the Ground.
- Signals A and B have two states, which are defined oppositely. For example, high value of signal A is known as "MARK" and low value is known as "SPACE", and high value of signal B is known as "SPACE" and low value is known as "MARK".
- The logic of SPACE is referred to as logic "0" and MARK is referred to as logic "1".

Universal Serial Bus (USB)

Universal Serial Bus precisely known as USB is one of the most popular industry standards dealing with the communication protocol, power supply, connecting interface including cable and connector used for the serial communication in the domain of telecommunication and computer networks nowadays across the globe. USB is extensively replacing many other standards of serial communications such as RS-232, RS-422, and many others. Almost all equipment used in computer networking is emerging with the capability of supporting USB communication and availability of the respective ports on the devices.

Multiple USB standards have been released till now. The last version of the USB standard in use is USB version 4 commonly referred to as USB4. The first version of USB was introduced in 1994. That version was named USB 0.7, and a couple of years later in 1996, the standard USB 1.0 version was released.[108]

The initial version was capable of speed just up to 12 Mbps at full speed, but normally, it was in between 1.5 and 12 Mbps. The advanced versions such as USB 2.0 and USB 3.0 were released in 2001 and 2011, respectively. The respective speeds of those devices were 480 Mbps (full speed) and 5 Gbps (full speed). The latest version USB 4.0 was released in 2019. This version is capable of handling speeds up to 40 Gbps, and all connections supporting this speed are known as SuperSpeed (SS) ports, connectors, protocols, or interfaces.

The governing body of the USB standards is known as USB Implementers Forum or precisely USB-IF. This forum has hundreds of members from across the computer, electronic, and telecommunication industries. There are many committees of this forum that deal with the different aspects of the USB protocols, connectors, cables, interfaces, and compliance. The most important committees of the forum include:

- USB Working Group
- Compliance Committee
- Marketing Committee

USB communication uses different types of communication ports and connectors, which are defined in terms of sizes and standard naming schemes. A few of them are listed in the following:

- Micro connectors
- Mini connectors
- A type connector
- B type connector
- C type connector
- SS supporting connectors

The main functions of USB communication protocol commonly desired/realized by the manufacturers and end-users include the following:

- Plug and play connectivity among computer peripherals
- Plug and play connectivity between other network nodes compatible with the USB protocol
- Providing power supply to the small peripherals through the host
- Storage of data on flash and other storage equipment
- Instant connection and disconnection for communication on-the-go
- Self-configuration and hot swappable features
- Monitoring activities on the serial bus

A few devices and connectors compatible with the USB communication protocol are shown in Figure 3.8.

The communication between the host and peripherals is controlled by the host, which generates all messages to control the communication among multiple devices. A host can communicate with 127 devices simultaneously with unique addresses, because USB communication supports 7-bit addressing scheme, which can accommodate up to 127 ports leaving the zero (0) slot for synchronization purposes in the communication.

USB communication supports star topology through two types of elements: pipes and endpoints. There are two types of pipes:

- Data pipes
- Control pipes

The control pipes are used for communicating the control of the transmission between host and devices while the data pipes are used for transferring the data between two

FIGURE 3.8   USB compatible devices and cables (Flickr).

entities. The data pipe and control pipes carry four different types of data transfers as mentioned in the following:[83]

- **Control transfers** – This transfer works on control pipes and is used for configuration, inquiries, and device commands.
- **Interrupt transfers** – These types of transfers are sent out on the data pipes and send the burst of data with guaranteed data and reduced latency.
- **Bulk transfers** – This transfer is done through data pipes in which the speed and latency is not guaranteed, but a huge amount of data is transferred.
- **Isochronous transfers** – They are done over the data pipes for transferring guaranteed data, speed, and latency without any error correction or resending of the packets.

There are two parts of host controller architecture that are used for different controls functions and capabilities. Those parts include:

- Controller part
- Root hub part

The USB host controller has three different types of controllers residing on the host control system, which are:

- Universal Host Controller Interface
- Open Host Controller Interface
- Extended Host Controller Interface

The communication based on control and data transfer between the host and the device takes place through different types of transactions to complete a data transfer activity. This communication transaction consists of the following packets:

- **Token Packet** – It is used for intimating about the next transaction that will follow.
- **Status Packet** – This packet is designed for error correction and acknowledgment.
- **Data Packet** – This packet contains the payload of the transaction.
- **Handshake Packets** – Used for connection establishment.
- **Start of frame (SOF) packet** – This packet is used for the synchronization of the host and device by sending a stream of an 11-bit frame.

A USB packet consists of different types of fields. Different types of packets have a different combination of fields to form a unique packet. All of those fields used in all types of USB transaction packets are mentioned in the following:[84]

- **Synch field** – This field consists of 8–32 bits. The purpose of this field in the packet is to synchronize the communication between two devices and inform the receiver that where PID bits will start from. The **Synch** field is available in all types of packets used in the USB communication protocol.
- **PID field** – This field is referred to as packet identification. The value of this field defines what type of packet is being sent or transferred. This is a 4-bit field to define different IDs of the USB communication packet. This field is always used in all types of USB communication packets used for USB protocol.
- **ADDR field** – This is the address field, which contains the address of the device that the message or packet is destined for. This field consists of 7 bits and can accommodate 127 addresses other than the zero position, which is left for other purposes. This field is available in data packets only no other type of packet has this field.
- **ENDP field** – This field is known as the endpoint field, which consists of 4 bits of space. It can accommodate 16 different types of endpoints. This field is used in only data packets no other type of packet has this field.
- **CRC field** – Cyclic redundant check (CRC) is a field that checks the errors in the data sent out to the destination device in the USB communication. This field varies in size. The data packets have a 16-bit field and the token packet has a 5-bit field. This field is available in data packets, token packets, and SOF packets.

- **EOP field** – End of packet or EOP field is used for notifying the device regarding the end of the packet stream. This is a 4-bit field, and it is available in all types of packets used in the USB communication protocols.

X.21 Protocol

This is another standard that defines the interface and communication protocol for serial communication between a DTE and DCE in digital communication systems. The connector used for X.21 serial communication protocol is known as D sub-connector No.15 or precisely, DB-15.

This protocol was introduced in majority of the European countries in 1980. This protocol was developed and released by the CCITT standard for serial communication. Different types and versions of this protocol were released for analog and digital communication during the past decades. The popularity of X.21 is decreasing with the advent of the other powerful serial communication protocols such as USB communication and others.[85]

X.21 protocol uses two types of connectors, which are used for circuit-switching circuit communication. They are:

- Balanced X.27 N.11
- Unbalanced X.26 N.10

The main features of the X.21 protocol are listed in the following:

- It is a state-driven serial communication protocol.
- It supports full-duplex communication.
- It supports data rates from 600 bps to 64 kbps.
- It supports circuit switching in public telephone networks by using synchronous ASCII codes.
- Call establishment was defined under CCITT standard (ITU) in "Blue Book" in 1988.
- This protocol uses seven types of different signals on different pin positions in the DB-15 connector for communication through the X.21 protocol:
  - **Ground Signal (G)** – This signal is used for conveying the logical state against the other circuits.
  - **DTE Common Return (Ga)** – This signal offers a reference for receivers. It is used in only un-balanced configuration of the X.26 protocol.
  - **Transit (T)** – This signal carries the payload or data from the DTE to DCE in three different phases – data transfer, call connect, and call disconnect phase.
  - **Receive (R)** – It is used for data transmission in the same three phases, but the direction is opposite – from DTE to DCE
  - **Control (C)** – Depicts the meaning of the data sent out to DCE from the DTE through this type of signal.

- **Indication (I)** – This signal is sent out to DTE in response to its control signal to indicate that data received on the receiving line.
- **Signal Element Timing (T)** – This signal is sent out by both DTE and DCE to each other for providing the information of sampling of the data.
- **Byte Timing (B)** – This is an 8-byte element timing signal sent out for indicating the transition of the circuit from ON to OFF state.

These seven types of signals are transmitted on predefined ports of the DB-15 connector. The three connectors are not assigned, and others are assigned with the aforementioned signals, which are governed by the X.21 serial communication protocol.

Sample Questions and Answers for Chapter 3

Q1. What are the three major techniques for digital telephone switching?

A1. Digital telephone switching can be divided into three major techniques as listed in the following:
- Space division switching
- Time division switching
- Time-Space division switching (hybrid)

Q2. In which switching technique, the entire message originating from a source is collected and stored in the memory storage?

A2. In message switching technique.

Q3. Name some of the most commonly used architectures for analog to digital conversion.

A3. The most commonly used architecture for analog to digital conversion are listed in the following:
- Successive approximation register (SAR)
- Flash/paralleled architecture
- Mesh architecture
- Sigma–delta (SD) architecture
- Pipelined architecture
- Delta-encoding (DE) architecture
- Intermediate FM stage architecture.

Q4. What is HART protocol?

A4. Highway Addressable Remote Transducer (HART) is a communication protocol that is used for transmitting digital signals over analog wiring systems among the Internet of Things (IoT) devices and the control systems. HART protocol was released at the beginning of the 1980s. This protocol was consistently getting grounds in the field of remote monitored devices.

Q5. What is a MODEM?

A5. The term MODEM is derived from two words: modulation and demodulation. The purpose of modem is to modulate the analog signals into digital signals and digital signals into analog signals.

PART TWO

Details of Digital Communication Protocols

Major Telephony Protocols

4

DIGITAL TELEPHONY COMMUNICATION PROTOCOLS

Analog telephony communication protocols were discussed in the previous chapters. Now, let us discuss the digital telephony communication protocols in this part. There are certain telephony protocols that were initially designed and developed for analog telephony systems, but later on, those protocols were modified and upgraded for digital telephony use.

We will talk about the important communication protocols used for telephone communication systems along with the narrowband data transmission.

Pulse Code Modulation (PCM)

Pulse code modulation, commonly referred to as PCM, is a type of modulation, which is extensively used in telephony systems across the world. It is extensively used in narrowband as well as in broadband data communication.

The PCM is the process of presenting the analog voice signals into the digital stream of zeros (0s) and ones (1s). The conversion of the analog signals into digital signals consists

DOI: 10.1201/9781003300908-6

of some unique signal processing activities. The analog signal is fed to the PCM converter electronic circuit, which performs the following three major processes on the analog signals and converts them into the digital stream of signals:[86]

- Sampling of signals
- Quantization of signals
- Coding of signals

The first step in creating a PCM-modulated signal is sampling of the analog signals. In this process, the input analog signal is sampled for its amplitude for a certain period of time. The sample of the analog signal is taken in terms of the amplitude of the analog signal at the time of taking the sample. If the sine wave of the analog signal is moving in the positive quadrant, the value of the sampled discrete signal will be positive and if the sine wave is traveling into the negative domain, the value of the sampled discrete signal will be negative. The value at the horizontal axis will be zero length of the digital discrete sample. For the analog signal, the samples are taken at certain periods of time as shown in Figure 4.1.

How many samples should be taken during a cycle of the sine wave is determined by the Nyquist sampling theorem, which will be discussed in the next topic in this chapter. The general rule of thumb is that the sampling rate is in direct proportion to the quality of the output signal. So, the greater the sampling rate, the better the quality of the output digital signal.

After sampling in the PCM, the second process is quantization of the sampled signals. The samples taken in the first step of the PCM will be measured in approximate values. The approximate values are taken because the sampled signals are discrete, but they vary

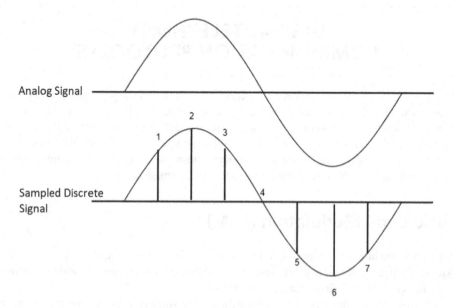

FIGURE 4.1   Schematic diagram of PCM sampling.

with the curve of the analog signal, which does not travel in discrete motion (rather that travels in the transient wave or motion).

The quantization measures the discrete sample of the signal in terms of voltages. For example, the first sample in the aforementioned figure measures 3.6 V and the second sampled signal measures 5 V, and so on. For the fifth, sixth, and seventh, the signal will measure the negative volts.

The quantization process also approximates the signal value by rounding off the fractional signals to the nearest approximate significant number. For example, the second sample will be approximated to 4 V because the nearest significant number of 3.6 is 4. Normally, these values are counted and considered up to many points after the decimal point.

After the process of quantization, the next step in the PCM is the coding of the quantized signals. In this process, the quantized values are divided into binary values. For example, the maximum signal value of 5 V can be coded into three binary digits. If the value of the signal is divided into even more precise values, then we can code using four binary digits to code the quantized signal into 15 discrete values in the binary system. For example, the value "4 volts" of the signal will be coded as 100, "3 volts" will be coded as 011, "2 volts" will be coded as 010, "1 volt" will be coded as 001, and so on. If every height of the sampled signal is measured in seven-digit binary system, then it will be able to cover 128 volts or any other equivalent units.

Nyquist Sampling Theorem

Nyquist sampling theorem is the fundamental principle of converting an analog signal into a digital signal for the PCM in telecommunication technology. The Nyquist theorem states that:

> Any analog signal can be converted into a digital signal through pulse code modulation by taking the discrete samples at equal interval either twice or higher times the maximum frequency of the analog signals. The signal taken as much as twice (min) the frequency of analog signal can be reconstructed into the analog signals without any loss of the information in the analog signal. This theorem also says that the greater number of samples would result in a better quality of the signal conversion.[87]

According to the Nyquist theorem, the voice signals can be sampled with 8,000 samples per frequency of the voice. The nominal frequency used for the voice transmission is 4 kHz. So, the nominal sampling rate for this frequency would be twice the frequency, i.e., 8,000 samples per second. The average time between two samples would be 1 second/8,000 samples. This will be equal to 125 microseconds. In PCM, 125 microseconds interval would be standard in different data streams.

The PCM modulation uses an 8-bit number for the coding system. Thus, it is a 64 kbps data stream commonly known as digital signal level 0 (DS0). These data stream units are combined together to form bigger data streams such as DS1, DS2, and DS3. The most common samplings used in the telecommunication transmission system for the PCM modulation include the following:

- E1 sampling stream
- T1 sampling stream

The stream samplings based on PCM are the combination of the PCM and multiplexing of the data streams of 64 kbps channels into two types of bundles (used in different parts of the world). These are known as E1 and T1 streams. Those streams use PCM at the single data channel level (64 kbps) and use the TDM technique to transmit different bundles of channels over a medium.

An E1 is a European standard for PCM technique for combining 32 channels of 64 kbps data rates. In 32 channels, the first channel (0 channels) is used for synchronization purposes, while the 16th channel is used for control functions. The remaining channels (30 channels) ranging from 1 through 15 and 17 through 32 are used for data and voice transmissions. The total capacity of an E1 stream is 2.048 Mbps, commonly referred to as 2 Mbps stream.

Similarly, the T1 is an American standard for the PCM and TDM of multiple channels into one bigger combo known as DS1. A T1 stream consists of 24 channels of 64 kbps data streams multiplexed in the time domain. The total capacity of the T1 carrier is 1.544 Mbps. It can simultaneously cater to 24 voice calls or 64 kbps data streams in a digital telephony system.

R1 Signaling for T1 Trunks

R1 signaling is an MF signaling system that uses multiple frequencies for regional telephony signaling. This was primarily developed for analog telephony systems but, later on, this was modified for digital telephony use.

R1 signaling is extensively used in the USA as a regional signaling system for North American countries. The detailed features, capabilities, and uses of the R1 signaling system have already been mentioned in Chapter 2. In the digital T1 network, the R1 signaling uses a bit-robbing technique for transmitting the line signaling as well as register signaling.

The bit-robbing is a technique for R1 signaling in the T1 transmission line. In this technique, every eighth bit of a frame is after every sixth frame. As we know, there are 24 channels or frames in a T1 carrier, so after every sixth frame, one bit is robbed. Thus, totally four bits are robbed in the T1 carrier. Those four bits are known as ABCD bits, which are used for the on-hook, off-hook, and other signaling through the CAS technique. The register signaling is done over the 64-kbps data channel for address and tones. The addresses include calling and called number.

R2 Signaling for E1 Trunks

R2 signaling is another form of MF signaling system, which was primarily devised for analog telephone signaling for both line signaling and register signaling. The details of this signaling have already been provided in the second chapter of this book. In this part, the details of its usage in digital E1 like basic rate interface (BRI) and primary rate interface (PRI) will be discussed.

This is the regional version of R1 MF signaling for the European countries. In this signaling, the 16 data stream in an E1 carrier is used for two types of signaling. The line signaling is used for the supervisory signals and the inter-register signaling is used for the call setup and control functions in an E1. In this signaling, only A and B bits are used for digital signaling in E1. For the analog signaling, all four bits, ABCD, were used for control signals. These bits are taken from the 16th frame of the E1 carrier.

Time-Division Multiplexing (TDM)

TDM was previously discussed in Chapter 1 while covering digital multiplexing. In fact, TDM is a type of multiplexing, which could be used in both analog and digital telephone systems for carrying voice as well as data. This is a technique in which multiple digital signals are combined together to form a data stream that can be used for transmitting bigger data at a higher rate. The examples of PCM coded E1 and T1 fall under the digital TDM commonly referred to as TDM.[88]

In this multiplexing scheme, the input digital signals are divided into equal blocks of digital signals and sent out over the equal intervals of the outgoing medium during an equal time duration known as time-slot. As we saw, there are 32 channels of E1 and 24 channels in a T1 carrier. Those channels carry data for a separate subscriber on a 64-kbps channel. The TDM technique is used in multiple transmission systems such as PCM, synchronous digital hierarchy (SDH), plesiochronous digital hierarchy (PDH), and synchronous optical networking (SONET). Those systems are based on both the electromagnetic signals used in copper and air interfaces as well as the light signals used in fiber optical transmission systems. TDM is one of the most popular techniques used in telecommunication transmission systems.

Digital Private Network Signaling Systems (DPNSS)

Digital Private Network Signaling System (DPNSS) is a proprietary communication protocol developed by British Telecom (BT). This communication protocol was developed and deployed in 1980. The main objective of developing this protocol was to provide dedicated ISDN links to corporate users. PRI service is a bundle of 30 channels of 64 kbits used for both data and voice services, while the two signaling channels commonly referred to as D-channels were used for signaling and control functions. The voice and data channels of a PRI trunk are referred to as B-channels of ISDN service.

Two versions of the DPNSS signaling system were developed. The first version was known as DPNSS 1, which is now not used anymore because the advanced version, DPNSS 2, has been designed. The DPNSS 2 would also be replaced by the latest standard protocol known as Q.931 for ISDN services over PSTN and other IP-based services across Europe.

The DPNSS 1 protocol was designed for interconnecting the PBX in a private network. The PBX was used by large organizations for voice and data services a few decades back. Nowadays, the software-based PBX has been introduced, which uses other versions of ISDN protocols for their respective communications.

The DPNSS 2 protocol was used to interconnect the PBX with the PSTN network for PRI services. The main features of DPNSS 1/DPNSS 2 are summarized as follows:[89]

- It is defined under specification ND1301:2001/3.
- The former specifications defined by BT are listed in BTNR 188.
- DPNSS 2 specifications are defined under BTNR 190.
- It is designed for digital trunk over E1.
- It is a type of CCS system.
- It is a subset of ISDN User Adaptation protocol.
- It is also referred to as DPNSS 1/2 User Adaptation.
- It is a proprietary protocol for communication of inter-PABX and PABX and PSTN.
- It is an OSI Layer 3 protocol.
- This protocol itself is divided into levels (not related to the OSI layers).
- Levels 1–6 are dedicated for the call setup and call tearing down.
- Levels 1–6 are compulsory for any PABX to be DPNSS compatible.
- The aforementioned levels are used for admin, telephone, supplementary services, and features.
- This protocol is a type of compelled protocol in which the request should get a response before sending the next message. If the message does not get a response in a certain time period, the resending of the message is initiated.
- The protocol messages are transported in the shape of International Alphabet (IA5) texts, which are later renamed International Reference Alphabet.
- It works over 2 Mbps links.
- It has a very poor synchronization known as plesiochronously synchronized.
- This protocol is also used in the hybrid VoIP PABXs by using an additional card for the DPNSS signaling system.
- The tunneling of DPNSS over the IP-based network is also possible nowadays.
- Both DPNSS 1 and 2 use data channels on the PRI for signaling transmission.
- There are three types of frames used in the DPNSS.
- The names of message frames include Unnumbered Information (UI) frame, Set Asynchronous Balance Mode Restriction (SABMR) frame, and Unnumbered Acknowledgment (UA).[90]
- Each frame has two versions – one is **Command** and the other one is **Response**.
- The signaling communication takes place in slot No. 16 of the 32-channel digital trunk.
- DPNSS is derived from Digital Access Signaling System 2 (DASS 2).
- DPNSS uses Link Access Protocol (LAP) at the data link layer for signaling communication.
- The physical layer of DPNSS is defined by the PCME1 interface.

I.421 Protocol

I.421 protocol is another advanced version of the Digital Access Signaling System DASS2. It is also known as the Euro ISDN protocol. The main objective of developing this standard signaling system is to establish a uniform network across Europe, which is compatible with all equipment in use. This protocol would replace the old versions of the digital trunk communication signaling system derived from DASS2 systems.[91]

This protocol uses coaxial cables for normal connectivity to the premises in the United Kingdom. The newer protocols and network connectivity media are also being used for the provision of the PRI and BRI services in the UK through I.421 protocol.

QSIG Protocol

QSIG is another very important telephone communication standard used for both traditional PSTN and VoIP telephony. This communication protocol is used for the communication between two digital PBXs operating within a Private Integrated Services Network commonly referred to as a PISN network. This protocol is also referred to as the ISDN protocol for PBXs for their intercommunication within a private network.

This communication protocol is an abbreviated form of Q Signaling (QSIG), which was initially developed by the European Computer Manufacturer Association. Later on, it was accepted and adopted by the European Telecommunication Standard Institute (ETSI) as an open-source standard to be used by any vendor who wants the interoperability of products in any private network.[92] The protocol was designed for ISDN services. At the connection level, the protocol is based on Q.931 standard communication protocol, which is used for setting up the basic call functions such as establishing a call, supervising, and terminating the call setup. On the other hand, the Remote Operation Service Element (ROSE) protocol is used for application-level communication. The ROSE protocol provides different interfaces for application-level communication such as X.500, X.400, and others, which will be discussed later.

QSIG protocol succeeded the previously adopted version of communication protocol known as Q.931/I.451. Presently, QSIG protocol is extensively used for both the traditional PSTN communication and VoIP communication for integrating the ISDN-based services through multiple PBXs in a private communication network environment. In a private network, (e.g., a large corporate network that consists of numerous offices located at different geographical areas connected through PBXs for communication among the offices), this protocol is used for establishing smooth and seamless communication between two or more PBXs. This is commonly referred to as PISN.

The functioning of the QSIG protocol is divided into two major layers – one is the basic call, BC layer, and the second is generic function, GF layer. The BC layer deals with the call set between two PBXs. This layer also ensures the smooth implementation of signaling between two PBXs of different vendors. The main functions covered by the BC layer include the following:

- Call setup function
- Call proceeding function
- Ring alert function
- Call Connecting function
- Call release function
- Call release complete notification

The GF layer handles the value-added additional services used on a private telephone network. The major events handled by the GF include the following:

- Caller Line Identification (CLI)
- Call waiting function
- Call diversion function
- Call intrusion function

The QSIG protocol is extensively used in modern private networks for different purposes; a few of them are listed in the following:

- Virtual private network (VPN)
- PSTN-based ISDN Services
- VoIP network
- Broadband private network
- PBX to equipment connectivity like voice recorder, router, mail, fax, and others

Let us summarize the entire QSIG protocol with the following points to remember about this protocol.

- It is a standard protocol adopted by ETSI for intercommunication between two or more PBXs in a private network.
- It can be used for both PSTN-based and VoIP-based PBXs in a private network.
- It is designed for ISDN services.
- It is the network layer protocol of the ISDN stack.
- The data link layer of ISDN is handled by the Q.921 specifications, which will be discussed later.
- The physical layer of the ISDN protocol stack is handled by the PRI and BRI interfaces, which are also specified by I.431 and I430, respectively.

Q.921 Protocol

Q.921 is a data link (layer 2) protocol in the ISDN communication protocol stack. It is a version of data link layer function referred to as Link Access Procedure on Channel D or Link Access Protocol D-Channel (LAPD) of the ISDN network. The Q.921 ISDN specification is based on the HDLC layer protocol.

This protocol offers a reliable service for the second layer in the ISDN protocol stack. The upper layer of the ISDN is QSIG or Q.931 signaling protocol as discussed in the aforementioned topic. This is an ITU standard for data link layer communication.[93] This specification is not an end-to-end communication between multiple PABXs across the network. This is a reliable communication protocol for Layer 2 through LAPD between two nodes of ISDN, which means that a network of multiple PABXs (that are not directly connected), cannot communicate through the Q.921 protocol. As we know, the SS7 communication protocol is an end-to-end protocol, which is able to communicate among multiple nodes in a network by routing and forwarding the messages from one node to the next one and so on.

We can summarize this protocol with the following major points for easy understanding:

- It is the Layer 2 or data link layer protocol in the ISDN protocol stack.
- It is an ITU standard used in multiple ISDN communication.
- It has two types of data link layer frames known as frame A and frame B.
- Frame A consists of 8 octets or 64 bits.
- Frame B consists of 8 octets or 64 bits fixed frame plus a variable length field, which carries load or information in the frame.

The frames of the Q.921 protocol carry different types of control and addressing information. The distribution of 8 octets of frame "A" is listed as follows:

- Flag 2 octets – Starting octet and ending octet
- Address high – One octet after the first flag octet
- Address low – One octet after the address-high octet
- Control 2 octets – After low address octet, two octets are control octets
- Forward Error Correction FEC 2 Octets – After control octets, 16 bits are designated for FEC (forward error correction) control

VOICE/DATA COMPRESSION TECHNIQUES

Compression is a very useful technique commonly adopted in the voice and data in telephone systems. This is also referred to as codec or coding/decoding and also compressor/de-compressor. There are different coding algorithms and laws used for voice coding. A few of those coding mechanisms are mentioned in the following.

G.711 Protocol

G.711 is a coding protocol designed for voice coding and decoding in telephony systems. It was developed in 1972 for use in the voice codec scheme of telephony systems. This was released in the name of PCM of voice frequencies in the same year. It is an ITU-T standard for *voice companding*. In telecommunication and signal processing, *companding* is a method of mitigating the detrimental effects of a channel with a limited dynamic range. The name *companding* is derived from the processes of compression and expansion of the voice signal. The main purpose of this protocol is to obtain high-quality voice at a very low frequency band like 64 kbps and others. This protocol was primarily designed for 64 kbps bandwidth for voice, and later, the same technique was also used for narrowband data communication.[94]

The main features of G.711 compression and decompression protocol are listed as follows:

- This algorithm uses audio signals of frequency ranging within 300 and 3,400 Hz.
- Sampling of the frequency is done at the rate of 8,000 samples per second.
- Uses 8 bits of value for quantization of the sampled signal.
- The tolerance of the sampling system is about 50 parts per million.
- 8-bit quantization for 8,000 samples per second makes a 64,000 bits per second channel.
- Two types of this protocol are commonly used in telephony systems – A-law and μ-law. Those two laws will be discussed in the next sections.
- It has two appendices known as Appendix I and Appendix II. The first one deals with PLC, while the second one deals with discontinuous transmission (DTX).
- The DTX deals with two important aspects of voice over the channel – voice activity detection algorithm and comfort noise generation CNG algorithm.
- This protocol is extensively used in the VoIP service for providing a better quality of voice.
- This is an uncompressed form of codec as compared to other techniques like G.729 standard, which will be described in the next sections.
- It offers very high quality voice calls and handling of multiple calls through this standard is very difficult while maintaining the normal quality of the call.
- G.711 has two types of codec schemes governed by the laws known as A-law and μ-law. Both of those laws are mentioned in the next topics.
- The latest releases like G.711.0 and G.711.1 are used for loss compression and narrow and wideband enhancement for data.

G.729 Protocol

This is another major codec (coder/decoder) system, which uses the compressor technique for providing efficient utilization of the available bandwidth. G.729 is a royalty-free protocol adopted by the ITU-T. This is used for both voice and data compression and extensively used for narrowband data services. Unlike G.711 codec system, this scheme cannot support DTMF signaling, high-quality audio, fax, and other similar types of services due to the compression technique it uses (to utilize the available bandwidth more efficiently, which may lead to degradation of the QoS). This is the reason why VoIP uses it very rarely (i.e., to avoid the degraded quality of voice).[95]

The main features of this codec protocol are listed in the following:

- This protocol is designed for 8 kbps bandwidth with compression technique as compared to the G.711, which is designed for 64 kbps channels.
- For speech coding/compression, it uses Code-Excited Linear Prediction and Speech Coding (CS-ACELP – which stands for "Conjugate-Structure Algebraic Code-Excited Linear Prediction") algorithm.
- It supports multiple extensions that provide different rates such as 6.4 kbps, 11.8 kbps, and others.

- It has been extended through multiple versions known under the combination of A and B.
- The extended features of this codec protocol include G.729A, G.729B, G.729AB, G.729.1, and others.
- The G.729B extension deals with silence suppression and G.729.1 deals with wideband speech and audio by using another algorithm known as modified discrete cosine transform (MDCT) code.

SIGNAL COMPRESSING LAWS USED IN TELEPHONY

The compressing or codec laws are used for the compression and decompression of the voice over the traditional digital data. The different techniques are used in different countries for the codec purpose, which is commonly referred to as the COMPANDING process. This term is derived from the two processes – compression and expansion. There are two major laws extensively used in telephony systems worldwide:

- μ-Law companding
- A-Law companding

Let us describe these laws with more details about their use in different countries and services.

μ-Law Companding

This companding law is based on an algorithm, which encodes the audio signal for the ISDN network in the telephony services. This uses a non-linear quantization of the signals to code the information for the efficient use of the digital bandwidth. According to μ-law, the low amplitude signals contain more information as compared to the high-amplitude signals. The main features of this law are given in the following:[96]

- This law is used in Japan and North American countries.
- The input signal is 14 bits and the sampled output signal is 8-bit codes.
- The value of μ ranges from 0 to 255 in the mathematical formula of this law.

A-Law Companding

This is another companding law used in the European countries as well as in other Asian and African countries. This is used in the E-1 transmission schemes, which are common in Asia and Europe. The main features of this codec law are given as follows:

- A is a constant factor in the mathematical expression, which qualifies this companding scheme as the A-law of companding.
- This scheme takes 13-bit input and produces 8-bit code as an output.
- This is extensively used to reduce the dynamic range of the signal.
- Better signal to distortion ratio.
- A-law is used as an internationally recognized companding scheme if any one of the countries uses this law.

Q.931 Protocol

Q.931 protocol deals with the signaling and call setup between two ISDN DTEs. It is an ITU-T standard, which was adopted in 1998. This protocol handles multiple information elements (IEs) specified for call setup, maintenance, and tearing down of the same. This protocol is also known as the connection control protocol for ISDN services.

The call setup in this protocol is handled through the multiple fields in the messages that exchange between the two DTEs. Those messages have different fields. The fields carry different types of information for the call set, maintenance of the connection, and termination of the connection. The first message used for the call setup in Q.931 protocol is the SETUP message, which is sent from the DTE to the network. This message has a bundle of information in it for the network to proceed with the call setup. The information includes dialed number, dialer's number, and many other IEs. This message also specifies what signaling system is being used for the network connectivity to provide detailed information to the DTE. In response to the SETUP message, the call proceeding message originates from the network. The call setup request gets mature after checking several parameters and conditions specified in the protocol. When the call proceeding response is received by the network element, it sends a ring-back tone to the dialing party and a ringing alert to the dialed party. When the dialed party picks up the phone, the *connect* message originates. The call starts after the connect message and the data/voice starts transmitting. When any one of the parties hooks on the telephone, the disconnect message originates from the DTE to the network. In response to the Disconnect, the Release message is issued by the network element. To complete the call, a response to the Release message is issued that is known as *Release Complete*. Thus, the entire procedure of call setup, maintenance, and tearing down of a telephone or data call establishment completes successfully.

The main features and capabilities of this protocol are listed in the following:[97]

- It is a transport protocol, which can be compared with the TCP in the TCP/IP protocol.
- Flow control is not supported by this protocol.
- No retransmission is supported.
- It works with reliable protocols at the lower layers.
- It uses the D channel for call control signaling.
- Multiple messages are used for call control in this protocol.
- Each message has multiple IEs such as display, bearer capability, channel identification, and many others.

- The number of IEs in each message is predefined and is not the same in all types of messages.
- A few messages of Q.931 protocol include alerting, call proceeding, connect, suspend, setup, resume, disconnect, and many others.
- All the commands or requests have response messages that also have specified IEs.
- The extended version of Q.931 is Q.2931 protocol, which is used for the broadband-ISDN and ATM network signaling.
- The extended version was not adopted extensively but used by a few switch manufacturers.

A Q.931 protocol frame consists of four major fields, which are listed in the following:

- **Protocol Discriminator (PD)** – This field shows what protocol has been used for the signaling purpose.
- **Call Reference Value (CR)** – The CR field identifies the calls that can exist at the same time. It is a real-time value to identify the connection.
- **Message Type (MT)** – This field specifies the type of message used for the call setup. A few important types of messages are mentioned in the aforementioned list.
- **IEs** – This field defines numerous types of information regarding the message, and each message has multiple specified numbers of IEs (as mentioned on the aforementioned list).

COMMON CHANNEL SIGNALING (CCS)

CCS is a type of signaling used in the telephony system. In this signaling system, the signaling channels are separate from the data channels. At the beginning of the telephony systems, the CAS schemes were in use, but with the improvements in the technologies related to telephony signaling, the concept of common channeling signaling soared high across the world.

The CCS, which was used extensively and is still in use, is known as SS7. There are other signaling systems that also qualify for the CCS systems. The major protocols of telephone that fall in this category are described in the following sections.

SIGNALING SYSTEM NO. 7 (SS7 PROTOCOL)

Signaling System No. 7 (mentioned previously in Chapter 1) is the most popular and very well-known CCS system. It is extensively used across the globe irrespective of the regions and countries. There are numerous variants of this signaling system used in the world. This signaling system is also referred to as many other names:

- CCS System No. 7 (CCSS7)
- CCS No. 7 (CCS7)
- Common Channel Interoffice Signaling No. 7 (CCIS7)
- Precisely, S7 and C7 signaling

This signaling system is used for call communication among the telephone network elements for managing phone calls, text messages, and billing of the services in the network. The roaming of the mobile networks and mobile number portability are also done through this signaling system. Both the local and international roaming are done through this communication protocol.

The mobile cellular networks use some variant of this signaling system. The separate signaling system has been implemented in the last decade for mobile number portability and universal number services in which the signaling gateways were developed for the signaling purpose in which all operators of the mobile services as well as fixed-line services were included for signaling purposes only so that the numbers of any network can be used with any other network seamlessly.

The most important services incorporated by the use of the CCS7 signaling system include the following:[98]

- Call setup, call management, and call teardown in the PSTN networks across the globe
- Personal communication system call setup
- Toll-free services
- Mobile cellular signaling system
- Number translation services
- Short message services SMS
- Number portability services
- International and local roaming services
- Prepaid billing services
- Value-added services for marketing and other purposes

CCS7 protocol is a standard protocol adopted by ITU under the telecommunication standardizing committee. It was first developed in 1975. After its first standard, many variants and versions have been adopted in the telecommunication sector. Before the advent of the concept of the CCS, the CAS was commonly used in the old telephone systems. The examples of those signaling systems included the different versions of MF signaling systems like MF R1, R2, and others. The major difference between the CAS and CCS signaling systems is that CAS works with the exchanges that are directly connected through physical trunks. On the other hand, the CCS signaling system does not require this pre-condition. A CCS network can communicate with each other without having direct trunks connected between two entities in the network. The signaling can communicate through separate signaling gateway or higher-level exchanges in the hierarchy of the telephone exchange systems such as local exchange, tandem exchange, transit exchange, and gateways.

There were so many limitations and downsides of CAS signaling systems, especially the inefficient use of the media and infeasibility of adding value-added services to those systems. To improve the performance of the signaling and effective use of the available telecom resources, the new system based on the CCS concept was introduced. Among CCS systems, SS7 signaling became the most popular and valuable for all operators across the world. The most common variant of the SS7 signaling system includes the following:

- American National Standards Institute (ANSI)/Bell Standard for North America
- ETSI standard for European and the like countries

An SS7 signaling system consists of three major nodes in the entire signaling system. They are listed as follows:

- Service Switching Point (SSP)
- Signaling Transfer Point (STP)
- Service Control Point (SCP)

The schematic diagram of an SS7 signaling system is shown in Figure 4.2.

In the schematic diagram of the SS7 signaling network (Figure 4.2), the end-users are normally connected with the service switching point or signaling points (SPs), which are normally the local exchanges that are connected to the SSPs. This entire network is connected through multiple signaling links that are referred to as types of links. The list of the signaling links that connect the nodes of the signaling system No. 7 elements is given in the following:

- **Type A** – Connects either SSP to STP or STP to SCP
- **Type B** – Connects two paired STPs with the other paired STPs
- **Type C** – Connects the paired STPs with each other
- **Type D** – Connects the paired STPs with other paired STPs but at higher hierarchical order. Types B and D are commonly referred to as B/D links.

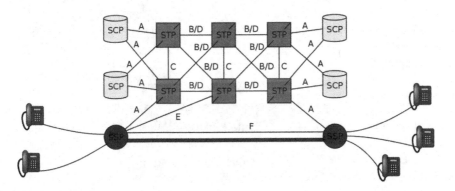

FIGURE 4.2 SS7 signaling network layout (Wiki Commons).

- **Type E** – Connects SSP to remote STP as an alternate route for redundancy and load sharing
- **Type F** – Connects SSP directly to the SSP if the presence of the STP is not available

All of the aforementioned types of links are shown in the schematic diagram of the SS7 signaling system. The signaling points are normally connected to the SSPs. The end-user telephones are connected to the SPs, which connect to the SSP for signaling purposes in an advanced IN based on the SS7 signaling system.

Signaling System No. 7 is a protocol that uses multiple protocols at different layers of the OSI model of communication. The full stack of an SS7 system is shown in Figure 4.3.

In Figure 4.3, you can see that there are three basic protocols MTP1, Message Transfer Part Level 2 (MTP2), and Message Transfer Part Level 3 (MTP3), which are common for all other upper-level services in this signaling system. Those basic protocols deal with the physical, data link, and network layers of the signaling communication among the network elements and the upper-level protocols deal with the services and applications at the upper layers in the OSI model of communication.

While the most important protocols are discussed separately in the following subsections, ASP stands here (in the figure) for "Application Service Part" and BISUP for "Broadband ISDN User Part" (BISUP). ASP provides the functions of Layers 4–6 of the OSI model. Currently, these functions are not required in the SS7 network and are under further study.

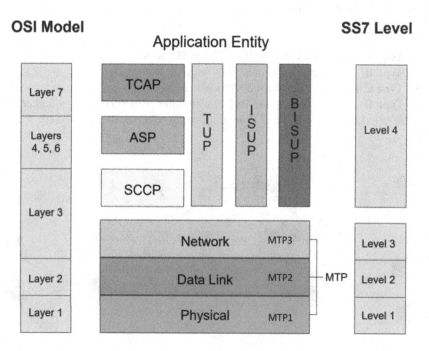

FIGURE 4.3   SS7 protocol stack.

The three basic lower-layer protocols MTP1, MTP2, and MTP3 of SS7 signaling are collectively referred to as the "message transfer part" (MTP) of the SS7 signaling protocol stack, and the upper levels of SS7 signaling protocols are known as "user parts" of the system. The three levels of MTP of the SS7 signaling can be compared with the first three layers of OSI models. It is very important to note that the SS7 signaling system was not designed in terms of the OSI model because both of those systems and concepts were developed almost during the same period and there was no clear collaboration between the two models in those days.

The OSI model uses the concept of layers, while the SS7 signaling uses the concept of level. The first three MTPs are referred to as level-1, level-2, and level-3 (see the figure). If we compare those levels with the OSI layers, we can say level-1 corresponds to the physical layer, level-2 corresponds to the data link layer, and level-3 corresponds to the network layer.

MTP1 Protocol

Message Transfer Part Level-1 or precisely MTP1 is the physical interface for the smooth transmission of the SS7 signaling. It is the first component of the basic three levels of transferring the messages across the nodes of SS7 signaling systems. The basic function of the MTP1 level is the smooth transmission of signal streams to the connected media between the two nodes. It defines the presentation of the signals on the physical media and delimiting of the signals for smooth and efficient transmission over the physical media.[99]

For delimiting, separation of the messages coming from the MTP2 is done through a flag, which is a stream of octets to let the opposite network element know that the next stream belongs to another message.

MTP2 Protocol

MTP2 is the second level of the S7 signaling system and the second component of the MTP stack of the system. It can be compared to the data link layer of the OSI model. At this level, the reliable delivery of the message is transported to the physical destination of the network. This level has three basic messages used for the communication, as mentioned in the following:

- Message Signal Unit (MSU)
- Link Status Signal Unit (LSSU)
- Fill in Signal Unit (FISU)

The main features and functions of the MTP2 protocol level of the SS7 signaling system are mentioned in the following list:[100]

- This is a type of narrowband signaling based on 56 or 64 kbps links.
- It uses variable sizes of messages referred to as signal unit (SU).
- It offers flow control, message sequence checking, error detection, and retransmission of unacknowledged messages.

- LSSU messages are sent to check the status of the link.
- If the link is up and there is no activity on the link between two nodes, the nodes sent out FISU messages, which are just information-less messages for letting the opposite element know that the node is alive.
- The information of the signaling is sent out through MSU.

MTP3 Protocol

MTP3 is the third-level component of the stack of message transfer part (MTP) in the SS7 signaling system. The functionality of this level can correspond with the network layer of the OSI model of communication. The level-3 part of MTP uses two major messages for performing the designed functions. Those messages are as follows:[101]

- Signaling Message Handling
- Signaling Network Management (SNM)

The main features and functions of this level of MTP stack are mentioned here:

- Proper routing of the signal message to the supported user application
- Taking self-healing technique in the case of the failure of signaling link in the network
- Uses point codes (PCs) as the addressing nodes such as telephone exchanges
- The originating exchange's unique PC is known as originating point code (OPC), and the address of the destination node is known as destination code point (DPC).
- Messages are routed based on the routing table manually created in the exchanges of the S7 communication system.
- SNM messages handle numerous network management functions like link status check, congestion check, monitoring link-sets and route-sets, and others.
- Three different variants of MTP3 exist in the world. They are ANSI variant, ITU-T variant, and Chinese GF variant.
- This is very important to note that MTP3 is a connection-oriented protocol that supports signaling based on the PCs, not the IP-based routing. So, the IP-based routing uses another format for communication in the signaling system.
- PCs are the unique numbers defined in the international perspectives. There are different formats of PCs based on different standards. For example, China and North America use 24-bit PCs in the format of three blocks of 8 bits. The ANSI standardizes this format of 24 bits. Japan uses 16-bit PCs, while the ITU standard defines 14-bit PCs. The ITU PC format is 3-bit:8-bit:3-bit.

TUP Protocol

The Telephone User Part precisely known as TUP is the first user-level protocol adopted in the SS7 signaling systems. This part works above the MTP3, which transfers the signaling messages to the desired destination. After the message has reached the destination node, the TUP part comes into play for the establishment of the telephone call connection between the two end-users.

This part of the protocol stack was developed for establishing, maintaining, and tearing down the analog telephone calls based on the circuit-switched signaling. This was extensively used for the POTS in the PSTN networks. It was very quickly replaced by its digital counterpart known as ISUP (Integrated Services Digital Network User Part). In many countries, the TUP was never used and ISUP was directly used in their digital services.[102]

TUP has a very limited support for any value-added features, which was the main bottleneck in the growth and popularity of this part of the protocol. It uses multiple messages for the setup, maintenance, and tear-down phases of the telephone call. This is very important to note that the messages related to TUP flow before the call is set up and the call starts, and then, just after the call ends. During a running call, no message travels between two nodes or end-users at this level.

ISUP Protocol

ISDN User Part precisely ISUP is another component of the "User Part" stack of the SS7 signaling system. It deals with the digital telephone and data services over ISDN digital network. It is the successor of the TUP, which had many drawbacks. To overcome those drawbacks, the ISUP was adopted to achieve better features and capabilities. This protocol is still in operation in the PSTN networks, where the ISDN services are used. It is also a major protocol that deals with the PSTN landline digital telephone services.

The main features and capabilities of the ISUP protocol are summarized with the precise points as mentioned in the following:[103]

- It is used in place of TUP to provide additional capabilities.
- It is used for establishing, maintaining, and tearing the telephone connection over ISDN services.
- The ISDN-based connection established through ISUP can be used for voice, data, fax, and other supported services.
- The ISUP message uses Circuit Identification Code (CIC), which is already specified between two exchanges, for establishing an end-to-end connection through multiple switches between two end-users on the ISDN network.
- The address of the dialed number is found through the routing table via ISUP messages, and the connection between end-users is identified by the CIC address among multiple exchanges to establish a connection.
- ISUP is also used for circuit status, management, and availability of the circuits between two exchanges.
- There are multiple ISUP variants used in the world such as ITU, ETSI, ANSI, and others.
- ISUP is capable of providing additional services such as billing control, call transfer, call waiting, CLI, and others.

All of the aforementioned features and capabilities are materialized through multiple messages used in the ISUP part of the SS7 protocol. Among those messages, a few very important ones are listed in the following with their respective functions.

- **Initial Address Message (IAM)** – This is the first message sent out by the ISUP that contains the dialed and dialing numbers along with the CIC information on which the connection can be established with that particular switch.
- **Subsequent Address Message (SAM)** – This is the second message if the entire number of the dialed entity is not completed in the IAM. This is a sub-set of the IAM message. This also provides additional numbers of the dialed entity. It may be single or multiple messages for sending out the dialed number in sequence.
- **Address Complete Message (ACM)** – This message is sent out from the terminating exchange when the address is complete, the dialed number is accessed, and the circuit from one end to the other end is seized. The originating exchange sends a ring-back tone to the dialing user.
- **Call Progress Message (CPG)** – The CPG message is also sent out by the terminating exchange to notify that all parameters of the call are normal and the call is ready to be answered.
- **Answer Message (ANM)** – This message is sent out by originating exchange to notify that the phone has been picked up. Now, the billing counter is started to record the talk-time. After this message, the ISUP as well as TUP parts remain silent in terms of sending out messages over the link until the disconnection occurs.
- **Release Message (REL)** – The "Release" message is initiated when any party in the conversation goes off-hook. A release message is also generated by the destination exchange, if the dialed number cannot be connected during the IAM dialing message. This is very important to note that every Release message has defined fields for the cause of release. The cause of release is sent out in the release message.
- **Release Complete Message (RLC)** – This message is generated in response to the release message generated by the exchange of any side. This is an acknowledgment that the call has been released and all resources have been freed for the others' use.
- **Connect Message (Con)** – The connect message is used for the automatic response machines or voice recorded numbers. This message is issued in response to the IAM message. This is not a normal call between two human end-users.

BISUP is an ATM protocol designed to support various services such as High-Definition Television (HDTV), multilingual TV, voice and image storage and retrieval, video conferencing, high-speed local area networks (LANs), and multimedia.

SCCP Protocol

Signaling Connection Control Part (SCCP) is a part of the SS7 protocol stack, which is used for intelligent services (IN) in the telecom networks such as PSTN and cellular mobile systems. This is a layer equivalent to the transport layer in the OSI model of communication. At the transport layer in the OSI model, the messages received through the network layer are routed to the service port number so that the right application can be accessed through the port number.[104]

In the SS7 signaling system, the routing through MTPs is done on the basis of PCs. When the signaling message reaches the MTP level-3, it needs to be routed to the right services or module of service. This routing is done by the SCCP protocol stack. This protocol uses sub-system numbers referred to as SSNs for routing the message to the right application or service named after the SSN. Similarly, the SSN of the Home Location Register (HLR) is 6 and Visitor Location Register (VLR) is 7. When a message for HLR is destined in the signaling message received at the MTP-3 level, the SCCP unfolds the SSN of the message and routes it to the HLR database for verification and other parameter checking purposes. If a signaling message contains the request for the pre-paid calling card query, the SCCP routes the message to the Service Control Point (SCP) of the IN network.

The SCCP layer protocol uses 256 numbers for Sub-System Number (SSN) to communicate with the upper-level applications in the SS7 signaling system. The numbers range from 0 to 255. Many of those numbers are already assigned for a certain service like SCP, HLR, VLR, SGSN (Serving GPRS Support Node), and many others and many of those numbers are also reserved for international and other services. A few numbers are also reserved for future use.

The main function of the SCCP protocol can be summarized as follows:

- Communicates a bridge between application-level services and MTP of the SS7 signaling system stack.
- Routes the message to the application level based on SSNs.
- Uses the functionality of Global Title Translation (GTT) to address and route the messages to the right node of service rather than checking the internal service level communication in the message. The GTT will be discussed in the following topic.
- Flow-control, message segmentation, protocol logic handling, and SCCP management and monitoring.

SCCP uses four types of message classes for the communication between the application and MTP3 on the network, which are listed in the following:

- **Class 0** – This class uses messages through a connection-less circuit (no-logical connection) and un-sequenced types of messages.
- **Class 1** – This message class deals with a sequenced stream of messages over the connection-less circuit (no-logical-connection).
- **Class 2** – This type of message class uses a connection-oriented circuit (logical connection) without any flow control capability.
- **Class 3** – This type of message class uses a connection-oriented circuit (logical connection) with the flow control capability.

Another very important work of SCCP is to check for the local and remote sub-system number. If an SCCP checks all the subsystem codes and then checks if that is local or remote, it will need a lot of processing time and resources. So, to expedite the process of routing at the SCCP level, the global title (GT) is used for immediate routing of the services in the signaling system network. The GT numbers are global codes that are translated into the PCs and vice versa. The destination and origination addresses received from the MTP level-3 messages are translated into GT and vice versa to make it easier for the SCCP protocol to choose the destination node instantly without peeping into the SSNs of the services.

The communication between the SSN and SCCP is done through the SCCP User Primitives. The user primitive is an interface established between the SCCP and the service defined with an SSN in the signaling system. There are two types of primitives used in this communication:

- Connection-less SCCP user primitives
- Connection-oriented user primitives

There are two classes of messages used in the user primitive communication over the connection-less communication. The responses to those messages are also used in this communication, but the core messages are those two. They include:

- **Unit Data (UDT) Message** – This is a simple unit of data used for the communication over connection-less user primitive.
- **Extended Unit Data (XUT) Message** – This is the extended version of the message, which is the bigger one and can be segmented over the connection-less user primitive interface.

The most important examples of services or applications using connection-less user primitive are Transaction Capabilities Application Part (TCAP) and ISUP.

Over the connection-oriented user primitive, different types of messages are used. First, a logical connection is established through certain messages, and then the data regarding the protocol, services, and GT are transmitted. The main messages used in the connection-oriented user primitive communication include the following:[105]

- **Connection Request (CR) Message** – This message is sent to the remote node for establishing an SCCP connection.
- **Connection Confirmation (CC) Message** – This is message is issued by the remote node in response to the connection request message.
- **Data Form 1 (DT1) Message** – This is the segments of data information transmitted over the connection-oriented user primitive.
- **Released (RLSD) Message** – This message is a notification to the SCCP for the connection release in the connection-oriented user primitive.
- **Release Complete (RLC) Message** – This message is used in the response to the RLSD message and used to confirm the release of the connection.

There are also some other messages that are used in the connection-oriented communication over user primitives such as connection refused (CREF), data form 2 (DT2), data acknowledgment (AK), expedited data (ED), reset request (RSR), and others.

Global Title Translation (GTT)

GTT is one of the SS7 protocols in which the SCCP translates the PCs of the nodes in the MSU message of the MTP-level 3 and translates it to the GT of the nodes both international and local in the signaling system. The entire process to modify the MSU for making it capable of routing with respect to the GTs is known as GTT.

This process is governed by a protocol that takes the data, especially the PC addresses of the originating and terminating points in the MSU message. In this case, a table known as GT Table is used. Describing this process in detail is beyond the scope of this book. This involves a software algorithm that uses the capabilities of the SCCP protocol to carry out this process of GTT.

TCAP Protocol

TCAP is a type of communication protocol over the SS7 signaling protocol stack. It is used for the communication between two nodes (machines) on the SS7 signaling network which have no circuit-oriented logical connection between them. This protocol is capable of handling multiple dialogues simultaneously within the same sub-system used over the SCCP protocol stack on the SS7 network. In this communication, the sub-system number is not used for simultaneous communication within the same sub-system, but the transaction IDs are used for handling the multiple streams of communication like a port number does in the TCP/IP protocol.[106]

TCAP protocol uses Abstract Syntax Notation One (ASN.1), which is a standard language for the description of a communication interface. This interface language uses different types of encryption methods in different applications. In the case of TCAP, the Basic Encoding Rule format of coding is used for defining the interface. The other coding schemes supported by the ASN.1 include Canonical Encoding Rule and Distinguished Encoding Rule. All these standards are defined by ITU-T under the recommendation named X.690. The main features and characteristics of TCAP protocol are as follows:

- TCAP is extensively used in the intelligent services powered by the application-level protocols named Intelligent Network Application Protocol (INAP) for fixed networks and Mobile Application Protocol (MAP) for mobile networks.
- TCAP handles two types of communication, i.e., messages and primitives. The messages are transmitted over the wire between the machines, while the primitives are sent over the local TCAP stack and applications.
- Basically, primitives and messages can be generalized as messages but all primitives don't qualify for the meaning of messages. So, they are differentiated as messages and primitives.
- Those primitives that transmit within the machines are not considered as messages rather they are called primitives.
- A primitive consists of either one or more than one part or component such as dialogue part and optional component.

- There are six main primitives defined by the ITU-T for TCAP communication. These include the following:
  - **Unidirectional OR Notice primitive** – This is the single and unstructured message/primitive, which can force transmission without Begin or End primitive/message in the communication.
  - **Begin primitive** – This primitive is sent out to start a dialogue with the remote TCAP user in the S7 signaling system. It has further primitives to follow.
  - **Continue primitive** – This primitive has further primitives to follow to continue the communication with the remote TCAP user based on the collected components of the previous primitives.
  - **End primitive** – This primitive is used to terminate the dialogue with the TCAP user. This primitive has no primitive to follow.
  - **Abort primitive** – This primitive is invoked when some error or unknown event has occurred. This primitive has no other message to follow, and the connection is terminated in the communication.
  - **Cancel primitive** – This is the only primitive, which does not qualify for a message. It sends in information to any node or application but stops the transaction after the counter of the timer reaches the maximum limit.
- Each primitive can have any one of the following component types:
  - Invoke
  - Return Result Last
  - Return Result Not Last
  - Return Error
  - Reject.
- The dialogue part of the primitive can have the following dialogues and uni-dialogue:
  - **AARQ** – used in the BEGIN primitive for IN services
  - **AARE** – used in END and continue primitives sent out in the response of AARQ
  - **ABRT** – used for the abortion of the dialogue.
- TCAP handles the communication based on two identification codes known as Transaction ID and Invoke ID.
- Transaction ID is used for communication between two machines, while the Invoke ID is used for operations of one transaction.

INAP Protocol

INAP is a communication protocol used for signaling between two nodes of an IN based on the ISDN fixed network commonly known as PSTN. INAP protocol is an ITU-T standard recognized across the globe for the realization of intelligent services in the traditional circuit switching network of telephones. The INAP protocol is defined under different compatibility sets. INAP protocol is used for the communication of three logical functions used in the INs, as listed in the following:

- Service Control Function (SCF)
- Service Switching Function (SSF)
- Specialized Resource Function (SRF)

The communication between those three functionalities is defined as compatibility set 1 (CS 1). The latest version of the INAP protocol is defined under CS 2. This set governs the communication between SCF and SSF and SCF and SRF. The main features, qualities, and characteristics of INAP protocol are listed in the following:[107]

- INAP is specified under the European Telecommunication Standard (EST 300 374–1).
- Like CAP (CAMEL Application Part) in mobile intelligent networking, the INAP also supports both Single Association and Multiple Association Control Functions referred to as Single Association Control Function (SACF)/Multiple Association Control Function (MACF).
- It is a part of the SS7 protocol stack used in the circuit switching networking.
- INAP protocol uses different addressing schemes such as PCs, sub-system numbers, and GTs.
- It works above the TCAP protocol in the SS7 signaling system.
- The fourth functional entity supported in the latest compatibility set is known as Service Data Function.[108]
- The protocol definition of the INAP protocol is defined into three sections, as listed in the following:
  - Definition of service primitives between two services
  - Definition of messages between two functional entities
  - Definition of actions taken at the respective entities.
- INAP runs different application service elements (ASEs), which are used for realizing the specific applications.
- In INAP CS1, 25 unique ASEs have been defined for different services.
- The ITU recommendation for INAP protocol is known as Q.1219.

V5 Protocol

V5 is an interface protocol for connecting the access network or last-mile network with the local exchange for providing PSTN as well as ISDN services. This protocol establishes the communication between the terminating equipment and local exchange for assigning the ports and type of service on the interface. The physical interface used in this protocol includes both fiber and copper connectivity, which are configured with the right version of the V5 protocol. The configuration of the V5 links is done on the basis of (E1) links.

There are two types of V5 protocols, which are described in the following list:[109]

- V 5.1 Protocol
- V 5.2 Protocol

V 5.1 protocol supports just one E1 of 32 channels of 64 kbps making the total link capacity equal to 2,048 kbps or 2.048 Mbps. The main features, capabilities, and characteristics of a V 5.1 protocol are as follows:

- V 5.1 protocol is specified by the G.703 specification.
- It supports one (E1) channel with BRI ISDN service.
- It does not support PRI ISDN service.
- The first channel is dedicated to the frame alignment.
- The 16th channel is always used for in-band signaling.
- It does not support concentration due to the unavailability of connection control protocol.
- All channels of the interface are C-channels and carry analog, digital basic, and other analog services without out-band signaling service.
- It does not support bearer channel connection (BCC) protocol.

On the other hand, the V 5.2 version can control from 2 to 16 E1 interfaces of 2,048 kbps capacity. It has many features and capabilities, which are different from the V 5.1 protocol. The main features and characteristics of the V 5.2 protocol are listed in the following:

- It is defined under G.704 specification.
- It can support from 2 to 16 E1 channels.
- Supports both BRI and PRI ISDN services.
- C-channels are protected from link failures.
- The failure of any link is switched over to the other link because a parallel protection protocol keeps functioning while operations in V 5.2 interface.
- Call basis connection is set up through BCC protocol running under the control of local exchange.
- Link ID verification is done through Layer 1 Finite State Machine (L1-FSM).
- Bit rates below 64 kbps are not supported at the control level.
- For the sake of additional communication security, V 5.2 interface uses additional reserved channels.
- V 5.2 protocol is also supported in VoIP communication.

Integrated Services Digital Network (ISDN)

Integrated Services Digital Network, precisely known as ISDN is an integrated service that is governed by a stack of protocols like the SS7 or TCP/IP. This is a comprehensive service that was designed for providing data and voice through the existing telecommunication network based on the copper wire network laid by the POTS providers. This service uses a modem for separating the voice and data for reception at the receiving end. The same modem is used for mixing the data and voice over the copper wire for the transmission.

This service uses a complete set of protocols that combine the SS7 signaling and TCP/IP network for the transmission of the data. The complete stack of the protocols is also referred to as the general-purpose protocol stack of services that offers numerous types of services over the existing PSTN network. The main services offered through the ISDN network include the following:

- Voice services
- Data services
- Text services
- Image services
- Video conferencing services
- Audio conferencing services

The ISDN service structure also provides the foundation framework for developing and establishing different types of modern telecommunication networks for future needs. This protocol stack is designed over the previous Integrated Digital Network, which was extensively used in the traditional circuit-switched network as a switching and transmission technology.

This is very important to note that the ISDN protocol stack is capable of supporting both types of telecommunication switching – packet switching and circuit switching – through different protocols at the network layer of the OSI model. The circuit switching at the network layer is handled by the Q.931 protocol, which was discussed before. The packet switching is handled by the X.25 protocol, which will be discussed later in this book. The protocol stack of ISDN service contains the following protocols:[110]

- BRI (I.430)
- PRI (I.431)
- LAPD (Q.921)
- Link Access Protocol Balanced LAPB
- Frame Relay
- Q.931 Call Control Protocol
- X.25 Packet Layer Protocol
- Application-specific end-to-end protocols

The ISDN service has capabilities for both channels – channel B and channel D to incorporate the suitable protocols for the provisioning of the aforementioned services through the traditional copper wire networks. The ISDN service is also known as the first version of broadband data cum voice services in the early 90s and beyond. The use of ISDN service is also used in some remote areas of the countries where modern data services have not yet reached. The complete protocol stack for D-channel and B-channel communication of ISDN services is shown in Figure 4.4.

ISDN services use three different types of channels to carry different types of information over the network. These three channels are as follows:

- Bearer channel (B)
- Data channel (D)
- Hybrid channel (H)

The capacity of every B-channel is 64 kbps. This channel is used for carrying information such as voice, audio, video, image, text, and data. The capacity of the data channel (D)

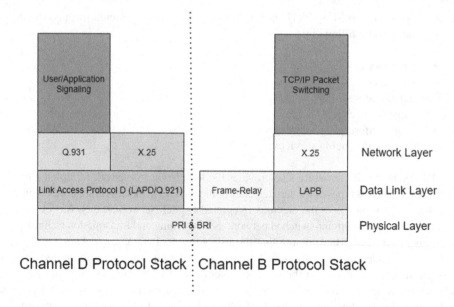

Channel D Protocol Stack : Channel B Protocol Stack

FIGURE 4.4   Protocol stack of ISDN service.

ranges from 16 to 64 kbps. This channel is used for call control, interruption, signaling, and other purposes. No bearer data flows in through this channel. This channel (D) is used for CCS purposes. It uses SS7, Frame Relay, and TCP/IP signaling at the upper layers.

The third channel used in ISDN services is the hybrid channel (H). This channel is a virtually combined channel consisting of two or more bearer channels that are used for multimedia services such as video, audio, video conferencing, and high-speed data transmission. With the advent of these ISDN services, the use of modern call conferencing became very popular among the public. The use of modern video conferencing applications also got traction after the advent of this service.

ISDN services are defined under two interfaces or electrical specifications are configured for a bunch of channels to carry the information. Those specifications are BRI and PRI. A BRI consists of two B channels known as bearer channels of 64 kbps each and one D channel known as a data channel of capacity 16 kbps. The schematic diagram of a BRI link is shown in Figure 4.5.

Thus, the total capacity of BRI for bearer services like voice, data, and other services is 128 kbps. This capacity can be either separated for connecting two devices with equal sharing of 64 kbps data or combined for one bigger channel. These channels carry the service data such as voice, video, audio, text, image, etc. The D channel is used for signaling and control functions of the BRI interfaces.

The electrical specifications of the BRI interface are explained in I.430 standard. The BRI service is also referred to as 2B+D service. In this notation, two B channels of 64 kbps each and one D channel of 16 kbps are mentioned. Thus, the total capacity of the BRI is 144 kbps.

FIGURE 4.5   Schematic presentation of BRI service (Flickr).

The PRI interface is an electrical specification governed by the I.431 standard. It consists of 23 bearer channels (B-channels) and one data channel (D-channel) of capacity 64 kbps. Thus, the total capacity of PRI is 1,544 kbps. This interface is also known as T-1 line in the USA and Canada. The capacity of the PRI used in Europe and Japan is different from that used in the USA and Canada.

There are 30 bearer channels and one data channel in PRIs used in the European standard. The total capacity of PRI in the European standard is 2,048 kbps. This PRI is referred to as the E-1 trunk. The E-1 trunk used as PRI is also referred to as the 30B+D notion.

DIGITAL SUBSCRIBER LINE (DSL)

DSL is also referred to as "Digital Subscriber Loop". It is a data and voice service based on modem technology. The modem technology uses the modulation and demodulation of analog signals over the existing copper-based twisted pair cable network. In other words, we can say that DSL is a physical layer data communication protocol that defines the transmission of the digital signals (data) through different techniques of modulation, mostly based on the FM.

DSL technology uses the following components for a complete set of services from the local telephone exchange to the subscriber premises.

- Digital Subscriber Line Access Multiplexer (DSLAM)
- Digital Telephone Exchange (DTE)
- Internet Server (IS) to connect to the cloud
- Local Loop Cable to carry the signal
- Splitter to separate the voice and data
- DSL Modem to demodulate the analog signals into digital
- R-323 or Ethernet Cable to connect the model to PC

A DSL service modem uses the 1 MHz bandwidth for the transmission of data at high speed. The frequency band is divided into two or three bands. The first band of 4 kHz is used for the telephone line or voice service, which runs on the same twisted cable multi-plexed by the DSLAM at the service provider end. The IS, which is definitely connected to the Internet cloud, provides the access to the Internet to the subscriber through DSLAM, which multiplexes the data over the upper band of the link.

The upper band is normally divided into two portions: one is reserved for the uplink data transmission, while the other one is dedicated to the downlink data. The capacity of the uplink and downlink data channels can be configured with different attributes to suit the different types of applications in data communication and modern telephony applica-tions. The image of a DSL modem is shown in Figure 4.6.

The DSL service is divided into multiple types of services based on the variation in uplink and downlink speeds, types of modulations used, type of multiplexing used, types of DSP schemes, and other factors. Based on those variations in the physical layer tech-niques and protocols, the DSL service can be divided into the following major categories:[111]

- Symmetric Digital Subscriber Line
- Asymmetric Digital Subscriber Line (ADSL)
- High-Speed Digital Subscriber Line (HDSL)
- Rate-Adoptive Digital Subscriber Line
- Very-High-Speed Digital Subscriber Line

All ADSL modems use discrete multi-tone (DMT) access technology. The DMT access technology is a variation of orthogonal frequency multiple access (OFDM). The modulation scheme used in the ADSL modems is mostly the QAM.

FIGURE 4.6   DSL modem (Flickr).

On the other hand, HDSL (High-Bit-Rate Digital Subscriber Line) modem uses the Carrierless Amplitude Phase (CAP) encoding with two-binary and one-quaternary line coding to send two bits in a symbol to enhance the speed of the DSL. Similarly, in the ADSL, different standard techniques such as DMT and non-standard schemes such as CAP and CAM (QAM) are also used.[112]

VOICE OVER IP (VOIP) TELEPHONY PROTOCOLS

Voice over IP, commonly known as VoIP telephony, is a packet switching-based telephony service implemented over the IP network. Our traditional telephone system, commonly known as POTS, works through circuit-switching systems under an integrated international telephony system referred to as PSTN.

The circuit switching is a connection-oriented switching in which a dedicated end-to-end connection is required to establish and continue a call during the entire period of call establishment and conversation and tearing down of the call connection. So, it is a very inefficient system due to continuously seized resources dedicated during the entire call accomplishment process. The VoIP telephone network connectivity integrated with the PSTN and cellular mobile networks is shown in Figure 4.7.

On the other hand, packet switching is much more efficient and connectionless. There is no need to establish a dedicated connection between two users during the conversation or even during the call setup. A packet of data originating from a dialing party can reach the dialed party through any suitable path, which is available without any congestion or any other disturbances. Thus, packet switching can use the resources, especially the media, which is

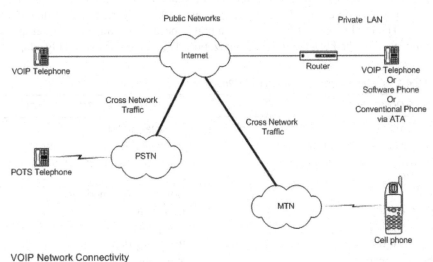

VOIP Network Connectivity

FIGURE 4.7   VoIP network connectivity (Flickr).

one of the costliest resources in the telecommunication field. The VoIP telephony can work easily on the public Internet, private Intranet, private LAN, and other IP-based environments.

VoIP is governed by two major signaling protocol stacks. Those signaling protocols use numerous other application-specific protocols to complete a stack of protocols for the VoIP telephony. The two signaling standards are as follows:[113]

- H.323 VoIP Signaling Standard
- SIP

The H.323 signaling standard is defined and recommended by the ITU-T. Again, the SIP protocol standard for VoIP is developed and standardized by the Internet Engineering Task Force (IETF).

The major components of the H.323 standard protocol stack are more than the SIP stack of VoIP telephony. The main components of the H.323 protocol stack include the following:

- Terminals
- Gateways
- Gatekeepers
- Multipoint control units (MCUs)

The main components of the SIP protocol stack are:

- User agents
- Network servers

If we compare both of the aforementioned two signaling and control protocols for VoIP, it will be very clear that the SIP protocol is much better than the H.323 in multiple aspects. But, still the use of H.323 in the existing market is more prevalent than the SIP protocol. The total market of H.323 is much higher because it came a bit earlier and was recommended by the ITU-T standard organization. This standard was quickly adopted by

TABLE 4.1   Comparison of SIP and H.323 Protocols.

SIP PROTOCOL STACK	H.323 PROTOCOL STACK
Simple protocol	Complex protocol
Textual representation for messages	Binary representation for messages
Full backward compatibility not required	Full backward compatibility required
Modular	Not very much modular
Highly scalable	Low scalable
Simple signaling	Complex signaling
Backed by IETF	Backed by ITU-T
37 headers only	Hundreds of elements
Easier Loop detection	Difficult loop detection

the big manufacturers of telecom equipment, and the second version of the same was also released within a short span of time.

The main reasons for the complexity of the H.323 stack of VoIP signaling protocol are that it was designed while keeping the ATM and ISDN in mind. The code presentation of the messages is binary in H.323 protocol, while the same is in text format in the SIP like the messages communicate in the HTTP environment. The major differences between the two VoIP telephony signaling protocols – H.323 and SIP – are described in Table 4.1.[114]

Let us now talk about these both protocols and their application-specific auxiliary protocols in the following sections.

Session Initiation Protocol (SIP)

The SIP is a VoIP protocol designed and approved by the IETF. This protocol is designed for establishing application-level control mechanism among multiple applications used in VoIP telephony. The major controls offered by this protocol at the application level include creating, modifying, and terminating sessions with different participant applications.

The working architecture of the SIP protocol is similar to the HTTP protocol, which uses a server and client approach in its communication. In this architecture, a client application originates a request to the desired server, which responds to the request with a particular response. The combination of request from the client and the response to that particular request from the server to the client is referred to as a communication transaction in this protocol.

The transaction messages are transmitted through INVITE and ACK messages to establish a channel for the communication of a transaction. The main features of the SIP protocol are mentioned in the following list:

- It is a fully reliable protocol without any dependence upon the underlying transport layer protocols such as TCP.
- For negotiation of the identification of codec techniques in the communication, SIP uses the Session Description Protocol (SDP).
- The SDP protocol also determines the media compatibility among the participants of the communication.
- It is a server–client-based protocol for VoIP communication.
- It is also capable of supporting the user mobility through proxy servers via redirect requests to find the location of the users.
- The SIP protocol provides the following functionalities in VoIP communication.
  - **Call handling** – This function contains the call transfer and call termination capabilities.
  - **User location tracing** – Traces the location of the end-user/system in the network.
  - **User availability tracing** – This function checks the availability and willingness of the user to participate in the communication.
  - **User capability agreements** – Identifies the type of media used between two participants and determines what media parameters to be used.

- **Call Setup** – This functionality includes the establishment of call parameters, ringing, and additional call-related features at both calling and called parties.
- The SIP protocol uses two types of components known as:
  - Network server
  - User agent.
- There are three components of the network server used in the SIP VoIP communication protocol. The names of those components are listed in the following:
  - **Registration server** – This server records and updates the current location of the end-user in the network.
  - **Proxy server** – This server receives the request for the user location and forwards the request to the next-hop server, which has more information about the user to locate the location of the server in the network.
  - **Redirect server** – This is another type of server that receives the request from the client for an address and finds the address of the next hop for processing the request and sends the address to the client from where the request was originated.
- There are two types of user agents (UAs). A UA is an end-system that works on the behalf of a user in the network. The names of UAs are as follows:
  - **User Agent Client (UAC)** – A UAC initiates the request.
  - **User Agent Server (UAS)** – A UAS receives and provides a response to the request.
- The communication in a VoIP communication system based on the SIP protocol uses numerous types of communication messages that travel within the network elements. A few important messages are listed in the following:
  - **INVITE** – It is used to invite a user for a call.
  - **ACK** – This message is used to acknowledge the reliable reception of communication messages in the system.
  - **REGISTER** – This message provides the location information of the user to the registration server in the VoIP network.
  - **OPTIONS** – This message is used to get call capability information regarding a particular VoIP telephone call.
  - **BYE** – This message is used for call termination between two parties.
  - **CANCEL** – The cancel message is used to terminate/abort the search for a user.
- The data within the message is presented with plain text commands not in the binary digits as is done in the H.323 protocol to be discussed next.
- The identification of a SIP host in the network is defined by a SIP URL. The URL is assigned to either a single user or a group of users.
- The address format of a SIP URL is → SIP:username@host.
- The unique user number is defined by the Call ID in the SIP messages.

The workflow of a SIP protocol-based call in VoIP network starts with an INVITE message from the client or calling party, referred to as "UAC or Client". This message is normally sent to the Proxy Server in the network. The proxy server tries to resolve the address and obtains the IP address of the called party and consults the location server for the location of the called party. This is important to note that the location server is not a non-SIP server. It stores the location information and updates the mobility of the users in the network. The location server provides information about the possible next-hop address for the location of the end-user. Once the IP address is resolved, the proxy server sends the INVITE request received from the client to the next-hop UAS. The UAS responds to the INVITE message through a response to the proxy server. The proxy server sends the response to the original client from which the INVITE message was received initially. The complete communication workflow is shown in Figure 4.8.

The second part of the communication is the acknowledgment process, which starts from the client with an ACK message to the proxy server, which acknowledges the UAS server for establishing communication between the two users directly.

The protocol stack of SIP-based VoIP telephony works with different types of protocols working at the fundamental three lower layers of the OSI model of communication, namely, physical layer, data link layer, and network layer. At the transport layer of the OSI model, this protocol stack uses both TCP and UDP for different processes used in VoIP communication.

For the transport and quality control functions, this protocol stack uses TCP protocol at the transport layer for reliable transmission of the control functions. The major protocols used for the transport and quality control functionalities include the following:

- **Real-Time Transport Protocol (RTP)** – This protocol handles real-time and time-sensitive data transmission in the system.
- **Resource Reserve Protocol (RSVP)** – This is a transport and upper layer protocol, which uses the TCP to maintain the QoS by reserving the network resources that ensure the desired quality of data over the VoIP network.
- **RTP Control Protocol (RTCP)** – This protocol offers control functionality over the RTP protocol for time-sensitive data to transmit over the network.
- **Session Announcement Protocol (SAP)** – This protocol works in tandem with the SDP protocol for announcing the multicast sessions.

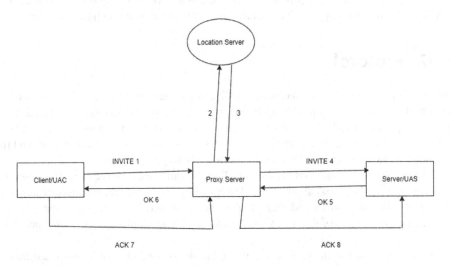

FIGURE 4.8 SIP workflow VoIP schematic diagram.

TABLE 4.2   Protocol Stack of SIP-based VoIP Telephony.

TRANSPORT AND QUALITY OF SERVICE			SIGNALING & CONTROL
			SIP
RTP	RTCP	RSVP	SAP/SDP
TRANSMISSION CONTROL PROTOCOL (TCP)			USER DATAGRAM PROTOCOL (UDP)
Network OSI Layer			
Data Link OSI Layer			
Physical OSI Layer			

- **SDP** – This protocol defines the parameters of the multimedia streaming and other time-sensitive multimedia data between two endpoints.

On the other hand, SIP is used in tandem with the SDP and SAP for the call signaling purpose. These protocols use the UDP transport layer protocols along with the TCP protocol at the transport layer for certain signaling and functionalities. The SIP-based protocol stack of VoIP telephony is given in Table 4.2.

This is very important to note that SIP was originally designed for establishing an interactive session for real-time communication over the newly evolving IP-based Internet network in the early 1990s. It is a so simple protocol that uses the simple plain text-based presentation of messages over the Internet like the messages of HTTP and Simple Mail Transfer Protocol (SMTP) used in the Internet communication over the Internet.[115]

The SIP protocol uses the similar technique that the HTTP uses for identification of the endpoint available on the Internet through Universal Resource Identifier (URI). A URI is located through a presentation known as a universal resource locator (URL). Another very important aspect of SIP protocol is its modular approach to accomplish communication through numerous application-specific protocols to integrate with the protocol.

H.323 Protocol

The H.323 protocol is a standard developed and adopted by ITU-T. This protocol was initially designed for multimedia conferencing purposes on the LANs assuming that the QoS is not provided in the LAN environment. Later on, this protocol was adopted for VoIP communication over IP-based networks. The first version was announced in 1996 and the enhanced version followed within a short period of two years in 1998.

This protocol was adopted by the ITU-T in the endurance for the seamless operability across the products and applications developed by a wide range of manufacturers and developers who adopt this standard in their respective products across the globe.

The salient features of the H.323 standard for VoIP communication are mentioned here:

- This protocol is an application-oriented protocol stack that runs over the transport layer of the TCP/IP network.

- It incorporates numerous application-specific protocols for different functionalities for handling the VoIP communication workflow.
- The H.323 protocol defines four logical components, which are listed in the following:
  - **Terminals** – The endpoints in the LAN system acting as clients are known as terminals. They can handle two-way communication in the network. A terminal in the VoIP network can communicate with its peer terminals, MCUs, and H.323 gateway element.
  - **Gateways** – The second logical component of VoIP communication based on H.323 is gateway. Multiple gateways can be available in a large network. This is the network endpoint, which can communicate with another gateway and H.323 terminals on IP as well as circuit switch-based networks. This component offers two-way communication in real time in the VoIP network. The translation among different formats of data and video codecs is also performed by the gateways while acting as a translator of the network. The connectivity between the Internet and PSTN network is also provided by the gateway components. In a direct communication within a LAN network, there is no need for any gateway.
  - **Gatekeeper** – This is a very vital component in the entire H.323 powered VoIP network. It acts as the controller and manager of the entire communication in terms of resources allocation, authorization, and call setup within its jurisdiction known as zone. A zone is the aggregation of all endpoints that are registered under its jurisdiction and the gatekeeper itself.
  - **MCUs** – The MCU is another important component of the VoIP communication network based on the H.323 protocol. It offers the capability of multipoint conferencing in the network among two or more than two endpoints (terminals) and gateways.
- All H.323 terminals are mandatorily supposed to support the following four major application-specific protocols:
  - **H.245 protocol** – Allows the channel usage
  - **Q.931 protocol** – Used for call set and signaling purpose
  - **Registration admission status (RAS)** – Used to communicate with gatekeepers
  - **Real-time transport protocol (RTP)** – Used to carry time-sensitive voice packets.
- A gateway is used for communication with terminals and gateways of the other networks. Thus, it should be capable of supporting two major protocols as listed in the following:
  - **H.245 protocol** – Allows the channel usage
  - **Q.931 protocol** – Used for call set and signaling purpose.
- The major functions of a gatekeeper are of very huge importance. They are listed in the following with proper explanation:
  - **Admission control** – The function to deny or accept the call on the basis of different criteria like source/destination addresses, authorization of called party, and many other additional criteria.

- **Call authorization** – For call authorization, gatekeeper uses H.225 protocol to reject the call from some terminals that have some kinds of restrictions on the terminals.
- **Address translation** – The resolution of the transport address against the alias address is done by the gatekeeper by using the registration table of the network.
- **Call signaling** – This is the responsibility of the gatekeeper to check and allow connecting call signaling channel directly in certain environments otherwise; it will choose the complete call signaling between two endpoints/terminals and gateways.
- **Call management** – This function handles the status of the calls so that it can detect the status of the destination terminal for efficient management of call processing.
- **Bandwidth management** – The rejection or acceptance of calls in certain conditions related to the bandwidth issues or congestions is also handled by the gatekeeper in the VoIP network powered by the H.323 protocol.
- The MCUs consist of two components as listed in the following:
  - **Multipoint controller (MC)** – This is a mandatory part of any MCU. It uses the H.245 protocol to determine the common capabilities of an endpoint/terminal.
  - **Multipoint processors** – This is an optional part. It is used for performing multiplexing of voice, data, and audio directly under the control of MC, which is a mandatory part of MCUs.

The call flow in the VoIP call set based on the H.323 protocol starts with the normal call set messages between the end-user and the H.323 powered gateways. In the recent advancement, the call connection process is based on the FastConnect procedure, which is adopted by the H.323 powered gateways to connect the call between two endpoints.

The calling party initiates the call set request to the gateway. The gateway senses the call request and initiates the call setup process by talking to the gatekeeper and other gateways over the H.225 protocol call setup messages. Once the call is verified for the establishment, the negotiation for the capabilities starts between the gateways over the H.245 protocol capability messages. Once the capabilities are negotiated, the media gateway information with all upper layer ports and channels is shared between the endpoints and other nodes of the network for further monitoring of the call over the IP network.

The schematic diagram of the VoIP network based on the H.323 VoIP telephone standard is shown in Figure 4.9.

The H.323 protocol stack uses an IP network stack at three lower layers of the OSI model. At the transport layer, it uses both UDP and TCP protocols for different applications. Those applications/functions include the transmission of data, signaling and control, audio and video, and registration. The following functions use the TCP protocol at the transport layer:

- Data transmission
- Control and signaling functions

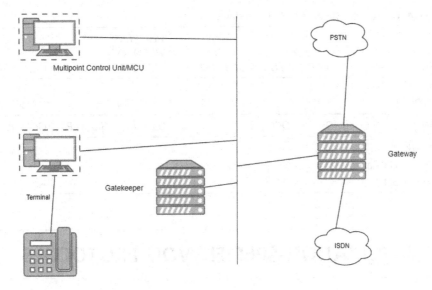

FIGURE 4.9   Schematic diagram of H.323-based VoIP network.

While the UDP protocol is used for the following functions/applications:

* Audio and video data
* Registration function

The upper layer protocol used for the data transmission is T.120 protocol. And for the control and signaling functionalities, the following two protocols are used:

* **H.225 protocol** – This protocol handles the call signaling in the H.323-powered VoIP system.
* **H.245 protocol** – It is used for the conference control functionalities.

For the real-time audio and video data transmission functionalities, H.323 system uses the following protocols:

* **Real-Time Transport Protocol (RTP)** – It is used for real-time audio and video data.
* **RTP Control Protocol (RTCP)** – This protocol is used for controlling the data flow and call participants that are using the RTP protocol for audio and video applications.

The complete protocol stack used in the H.323 protocol stack is shown in Table 4.3.

The registration of the user is done through H.225 Registration, Admission, and Status (RAS) protocol. This protocol is used between gatekeepers and gateways of the VoIP systems based on H.323 to register, admission, and status checking purposes.

TABLE 4.3   Protocol Stack of H.323 Standard.

DATA	SIGNALING AND CONTROL FUNCTIONS		REAL-TIME VIDEO AND AUDIO	REGISTRATION
T.120 PROTOCOL	H.225	H.245	RTP/RTCP	H.225 RAS
TRANSMISSION CONTROL PROTOCOL (TCP)			USER DATAGRAM PROTOCOL (UDP)	
Network OSI Layer				
Data Link OSI Layer				
Physical OSI Layer				

APPLICATION-SPECIFIC VOIP PROTOCOLS

VoIP telephony is governed by two major protocol standards – H.323 and SIP, which consist of numerous application-specific protocols that handle different processes used in VoIP calls. Those application-specific protocols work with the major VoIP protocol standards to complete the stack for VoIP telephony. A few of those protocols are described in the following.

Session Description Protocol (SDP)

The SDP is a protocol that works with the SIP protocol for the description of the capabilities of the endpoint. This protocol is normally used at the time of session negotiation between two endpoints (users). This protocol is used for control and capability establishment functions. It does not transfer the real data but only control data before the establishment of the session.

The SDP protocol is used for negotiating the conditions and capabilities for the transmission of the multimedia content, which is time sensitive and requires quality and ample resources for the same. The message of SDP is normally encapsulated within the SIP protocol to inform the other party regarding the capabilities and ways to communication with that particular endpoint in the system.[116]

The SDP message informs the other endpoint regarding the following capabilities, status, and conditions for transmission of media:

- The IP address of the endpoint, which will receive the data over the VoIP telephony communication network
- Declares the port number for receiving the desired stream
- Informs about the capability of the endpoint that what types of codec can support the endpoint
- Also, informs about the capability of codec in terms of mode communication such as capable of sending only or sending and receiving

H.245 Protocol

H.245 protocol is a control protocol for the multimedia communication of the VoIP telephony powered by the H.323 communication standard. This protocol is used to establish a logical control channel between two endpoints or users. The signaling takes place with the signaling entities located at each endpoint. The signaling entities are located as the incoming signaling entity, which is located at the called party and the outgoing signaling entity is located at the called party. Both of those entities are logical components, which are activated in their respective positions of their endpoints.

The call signaling is done through the H.225 protocol, and the H.245 negotiates the capabilities of the endpoints – calling and called parties, after the completion of the call signaling. The negotiation of the H.245 protocol is done through multiple messages. The negotiation of capabilities is established to control the multimedia data over the VoIP system based on H.323. The media stream is controlled on the basis of the logical channel and information shared by the H.245 protocol.

The major functions and activities that are performed by the H.245 communication protocol are mentioned in the following:[117]

- Negotiating and agreeing upon the media formats and the bandwidth for data transmission between the two endpoints
- Multiplexing of multiple media streams for the multimedia service
- Defining master and slave status of the endpoints

This is important to note that H.245 is a controlling protocol; hence, the transmission of data is not handled by this protocol, but it just establishes a controlled channel to negotiate and establish a multimedia channel in the communication.

It uses the following message groups for the communication between two parties in the communication.

- **Request** message – to ask the recipient to respond for an action
- **Response** message – sent out in the reply to the request message
- **Commands** message – It is sent to instruct the entity to perform some action but no need to send a response
- **Indications** message – This message group is used for sending information. Neither response nor action is required.

H.225 Protocol

The H.225 is a call signaling protocol used in the VoIP telephony under the H.323 standard launched by the ITU-T section. This protocol is a very important part for the transmission of multimedia, voice, audio, and other time-sensitive data. This is a call controlling protocol, which transports the messages over the Q.931 protocol. The main features and functionalities of the H.225 control protocol systems are listed in the following:[118]

- Establishment and termination of VoIP call over H.322 standard
- Supports the status inquiry capabilities on the endpoints

- Establishing ad hoc multipoint call expansion
- Supports call forwarding and transferring with a limited capability
- The message of H.225 is transported over the Q.931 protocol as a component of the user element information field of Q.931 protocol
- Defines how the packet, data, video, audio, and control information can be managed and controlled
- This protocol has two major components – H.225 RAS and Call Control
- A few basic messages used in the establishment of multimedia connections through the H.225 protocol are listed in the following:
  - Setup message
  - Call proceeding message
  - Alerting message
  - Connect message
  - Notify message
  - Release complete message

T.120 Protocol

T.120 is a series of different protocols for point-to-point and multipoint multimedia communication over the IP, ATM, and LAN network environments. This series of protocols is defined by a committee of ITU. The name has also been drawn from the name of that committee. This entire set of protocols is defined for the implementation of multipoint conferring services in the VoIP environment based on the H.323 protocol stacks.[119]

This protocol is used for the following services:

- Point-to-multipoint communication between VoIP-enabled devices
- Teleconferencing services
- Video conferencing services
- Computer-supported collaboration services
- Online chatting services
- File sharing and app sharing

Examples of the use of this protocol include WebEx app, NetMeeting, and others. There are many other specific versions of this protocol suite or protocol stack. A few of them are listed in the following:

- **T.121 protocol** – This protocol handles generic application templates.
- **T.122 protocol** – Specifies the definition of multipoint communication service.
- **T.123 protocol** – It is the transport layer (OSI Layer 4) protocol for the conferencing application on data networks. It is used as an error detection and correction part of the conferencing services. It supports underlying ATM and IP networks.
- **T.124 protocol** – This protocol is used for the generic conference control (GCC) function.
- **T.125 protocol** – Specifies the multipoint service (MCS) layer protocol specifications and works in tandem with the T.122 protocol.

- **T.126 protocol** – This protocol is defined for the still image and annotation transmission.
- **T.127 protocol** – This protocol is used for the transfer of files in binary format.
- **T.128 protocol** – The application sharing is handled by this protocol.
- **T.134 protocol** – This protocol handles text chatting in an application.
- **T.135 protocol** – It controls the user-to-reservation transactions in the system.
- **T.137 protocol** – This protocol is used for establishing and managing the control of the applications on a remote device.

Real-Time Transport Protocol (RTP)

Real-Time Transport Protocol, precisely referred to as RTP, is a real-time audio and video transmission protocol extensively used in the VoIP and multimedia application environments. This protocol is used in both major standards of VoIP telephony services – SIP and H.323-based telephony – for transferring the real-time voice and video data. This protocol was developed by the IETF and released in 1996 under the specification named RFC 1889. The second version of the standard was released in 2003 under the recommendation known as RFC 3550. The main characteristics of the RTP protocol are listed in the following:[120]

- RTP uses the connection-less unreliable UDP protocol at the transport layer for faster delivery of packets without any packet retransmission.
- It does not support multicasting or port number mechanism.
- It uses sequence numbers and time stamps for sequencing and real-time ordering.
- RTP supports multiple formats and codecs for the audio and video data such as MJPEG, MPEG, and many others.
- Being a real-time data handling protocol, RTP is very sensitive to packet delays but less sensitive to packet losses.
- It is extensively used in VoIP telephony applications.
- Other applications of this protocol include video conferencing, audio conferencing, video streaming, and audio streaming over the Internet.

It is very important to note that the RTP protocol establishes, transmits, and tears down the real-time session in collaboration with the RTP Control Protocol, commonly known as the RTCP protocol, which is explained at length in the next topic.

The RTP packet structure of the RTP protocol is shown in Figure 4.10, which specifies the different fields and length of the fields used in the RTP protocol packets.

Let us explain the different fields used in the RTP packet for real-time audio and video data communication.

- **Ver** field – It is a two-bit field, which defines the version of the RTP protocol.
- **P** field – This field is used for providing information about the presence of padding in the packet. It is just a one-bit field. If the bit is zero, that means padding is not present in the packet. If the padding field is 1, that means the padding is available.

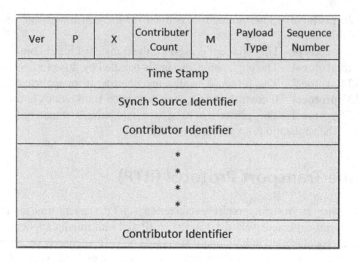

Ver	P	X	Contributer Count	M	Payload Type	Sequence Number
Time Stamp						
Synch Source Identifier						
Contributor Identifier						
* * * *						
Contributor Identifier						

FIGURE 4.10   RTP packet description.

- **X** field – This field specifies the extension of the header. That means, if the field extension value is 1, then the extra extension header between the basic header and data is available. If the value of this field is zero, which means the extension is not available. This is a one-bit field.
- **Contributor count** field – This field is defined for the number of contributors in the conferencing or sharing. Normally, this field is a 4-bit field, which means that the total number of contributors supported by this field is up to 15 numbers.
- **M** field – The M field denotes the marker field. This field is also a one-bit field. This field is used to indicate the end-marker of the data in any application.
- **Payload type** field – This field is used to describe the type of payload. This field is a 7-bit field, which can specify up to 128 types. In this field, different codes have been designated for different formats of the payload. A few examples of such format numbers are as follows:
  - Field value 0 = PCM Micro Audio
  - Field value 2 = G721 Audio
  - Field value 32 = MPEG1 Video
  - Field value 33 = MPEG2 Video
  - And many others
- **Sequence number** field – The sequence number field specifies the sequential number of the RTP packet to maintain a sequence of the packets traveling to the destination. This is a 16-bit field to name the packets for a session. The sequence number for the first packet is selected randomly, and then the sequence number continuously increases by one after every packet. The sequence number field is very useful in finding out the lost packet by mismatch rearrangement of the packets at the receiving end.
- **Time stamp** field – It is a 32-bit field. It is used for calculating the time of the packet. The time stamp for the first packet is given randomly like in the sequence number field. After that, the time stamp is the sum of the previous

packet and the time required to create the first byte of the packet. This series continues for the entire session.

- **Synch source identifier** field – The synchronization source identifier field is a 32-bit field. It is used for the identification of the source. It is a random number given to the source that initiates the conferencing or media streaming in an application. In very rare cases, when the sequence of the streaming of a packet is the same, this field helps resolve the conflict by the unique number chosen by the source itself randomly.
- **Contributor identifier** field – This is a 32-bit field. It is used for the identification of the source contributor. This is a unique ID allocated to the 15 unique contributors because the packet has a limit to define 15 unique contributors in the contributor count field.

Real-Time Transport Control Protocol (RTCP)

The Real-Time Transport Control Protocol (RTCP) is a sister protocol of RTP protocol. It is used for the control and statistical information sharing and feedback in a multimedia streaming session. It does not transmit the multimedia data but gets feedback on the quality and packet losses in the session running under RTP protocol in any VoIP telephony system. It uses the underlying protocols at the transport layer and RTP protocols.

This protocol sends out packets periodically to all contributors in a session to get feedback about the quality of service (QoS) through getting statistics of the packets received, lost, and other information in the live session. On the basis of the feedback received from the participant entities, the correction and modification are made to improve the QoS by changing the codec systems and flow of the data. The main characteristics of the RTCP protocol are mentioned in the following:[121]

- It is used in partnership with the RTP protocol.
- It does not transmit the payload or data in the packet.
- It gets feedback about the QoS by getting information about the packet loss, round-trip delay, packet counts, variation in packet delays, and other information.
- The RTCP packet is sent out on the odd-number port number of the UDP and then the even number used by the RTP protocol for data transmission.
- The RTCP reports are sent out by every participant in the session.
- The usage of bandwidth for RTCP communication should not exceed 5% of the total bandwidth of the session.
- The minimum time for multimedia participants to send the RTCP report is 5 seconds. It may be configured for a higher time too.

The header of the packet of RTCP protocol with all fields is shown in Figure 4.11. The details of the fields and their objectives are mentioned here as well.

- **Version** field – This field is used to specify the version of the RTCP. The latest version of RTCP/RTP is version 2. Version 2 is recommended under RFC 3550. It is a 2-bit field.

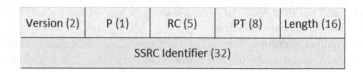

Version (2)	P (1)	RC (5)	PT (8)	Length (16)
SSRC Identifier (32)				

FIGURE 4.11   RTCP packet header.

- **P** field – This field is used to indicate whether padding is available in the packet or not. This is a 1-bit field as shown in the figure.
- **RC** field – This field is named reception report count (RC). It is a 5-bit field, which denotes the reception report blocks contained in the packet.
- **PT** field – This is a packet type field, which is an 8-bit field. This field specifies the type of the RTCP packet in the communication.
- **Length** field – The length field specifies the total length of the RTCP packet including the data or information in it. It is a 16-bit field.
- **SSRC** field – The SSRC is the short form of synchronization source identifier. This field is of 32-bit length. It is used to identify the unique source of the stream.

The RTCP protocol uses different types of messages in its communication with the participants of the media streaming session. The most important messages used in the RTCP are listed in the following:

- **Sender Report** (SR) – Sent out periodically after a fixed internal equal or longer than 5 seconds to inform about the quality-of-service statistics.
- **Receiver Report** (RR) – This report is sent out by the passive participants in the streaming to inform them about the QoS it receives.
- **Source Description** (SDES) – This message is sent out after a fixed interval of time to provide additional information about the participant parameters.
- **Goodbye** (BYE) – This message is sent out by the source that shuts the streaming in the communication.
- **Application Specific** (APP) – This message is used for designing the application-specific communication in the streaming.

Inter-Asterisk Exchange (IAX)

Inter-Asterisk Exchange (IAX) is an open-source protocol for VoIP communication with reduced use of bandwidth. This protocol was created by Mark Spencer, and the additional development was carried out by the open-source community. The first version of IAX has been superseded by the second version, which is commonly referred to as IAX2.

This protocol was primarily designed for faster transporting of VoIP communication sessions, especially the voice sessions between the endpoints and servers. This protocol can be compared with other major protocols like H.323, SIP, MGCP, and others. But, the use of this protocol is limited to some types of soft-switches and smartphones powered by

VoIP telephony. The main features and characteristics of the IAX2 protocol are listed in the following:[122]

- It was released in 2010 under informational or non-standard-track.
- It uses the UDP port for communication of both signaling and data simultaneously on one single port, which is fully different from the other VoIP protocols.
- The IAX2 uses the UDP port number 4569 for the communication.
- This protocol is a binary-encoded protocol.
- It supports multiplexing and trunking functionalities on the same channel.

Skype Protocol

Skype protocol is the first VoIP protocol, which was launched in 2003. It was launched by a startup company, which was acquired by Microsoft Corporation. Nowadays, it is an application owned by Microsoft Corporation (at the time of writing this book). It is a closed-source and proprietary protocol, which has not been made public for study. Even reverse engineering to decode the basics of the protocol is also prohibited under business laws.

The first version of the Skype protocol was deprecated by Microsoft in 2014, which is not accessible nowadays to log in through a simple client software. The new version of Skype protocol supports numerous additional features and capabilities. Numerous changes and advancements in the new version were introduced. The most important capabilities introduced in the latest version are security and privacy through strong encryption and security protocols. The schematic diagram of the basic structure with all components of a Skype network is shown in Figure 4.12.

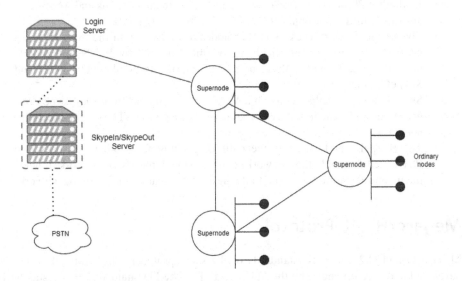

FIGURE 4.12 Skype network schematic diagram.

The main features and characteristics of the Skype protocol that we know are listed in the following:[123]

- It is the first peer-to-peer protocol in the field of VoIP technology.
- The network consists of three major entities of the network as given in the following:
  - **Login server** – This is a centralized authorization and authentication server.
  - **Super node** – This is a proxy server that handles the traffic and connections of the communications under a small group. It also works as a relaying agent in a semi-distributed network.
  - **Ordinary node** – A user or client node is known as an ordinary node in this protocol.
- Each client node of the skype has stored information about their respective super nodes, IP addresses, and the port number for reaching the super node in the network.
- Skype protocol works smoothly, efficiently, and without any restrictions behind the network firewalls and network address translation protocol.
- Before 2012, any computer with sufficient processing capacity could become a super node of the network, but after 2012, the policy to allow private super nodes has been reversed, and now, all super nodes are Microsoft's own servers located in their data centers at different parts of the world.
- The Skype protocol supports only IPv4 protocols and does not support IPv6 protocol in its VoIP communication network.
- Skype protocol integrates the Microsoft notification protocol for its chatting/messaging applications, and for messaging and signaling encryption, it uses the RC4 encryption technique.
- For voice encryption, Skype protocol uses the Advanced Encryption Standard (AES) to encrypt the communication from end to end.
- It allows API for integration of skype into the web applications and websites or any other third-party application through white paging only.
- The Skype protocol uses two other nodes to connect with the external voice networks and traditional telephone systems powered by PSTN and mobile networks. The names of those nodes are referred to as SkypeOut Node and SkypeIN node.
- SkypeOut server helps establish a VoIP call originating from the Skype ordinary node and route it to the external networks such as PSTN and other cellular networks.
- The SkypeIN node handles the calls originating from other networks like PSTN, cellular, and other networks connected with the Skype network. It helps the incoming call to terminate to the desired destination in the Skype network.

Megaco/H.248 Protocol

MEGACO and H.248 are the standards of the same protocol. This protocol is jointly developed and recommended by the ITU and IETF. The ITU named it H.248 standard, and the IETF refers it to as MEGACO standard. It is used for the call controlling between two telephony systems known as PSTN and IP telephony. It is also referred to as Media Gateway Control Protocol in the soft-switches used for the class-5 (C-5) telephone network in the integrated telephone communication systems.

It is an application-specific protocol, which controls the communication between media gateways, precisely referred to as MGW and the media gateway controller referred to as MGC. These two entities are two virtual modules in the IP-based telephony system working with C-5 soft-switches in telephone networks. This protocol used in Universal Mobile Telecommunication System (UMTS) network powered by the IP-based communication applies on the "Mc" interface of the UMTS network.

H.248 or MEGACO protocol can be implemented within two major backbone technologies such as IP network and ATM network in core UMTS networks or in any other SS7 signaling-based systems. This protocol uses TCP, UDP, and SCTP protocols at the transport layer and can directly establish communication between these protocols with H.248 protocol at the application level.[124]

In the ATM core network environment, the H.248 protocol communicates with the MTP3B protocol, which communicates with Signaling ATM Adaptation Layer protocol used in the ATM protocol stack. The communication at the application layer between media gateway (MGW) and media gateway controller (MGC) is based on the master/slave network communication in the H.248 protocol. This communication is extensively adopted in the IP-PSTN communication at the edge of the packet networks used in VoIP telephony. The schematic description of the H.248/MEGACO protocol stack working with the underneath protocols is shown in Figure 4.13.

The most salient features, characteristics, and capabilities of the H.248/MEGACO protocol are mentioned in the following list.[125]

- H.248 is a master/slave communication protocol between two parts of a soft-switch layer to handle the IP-based calls between PSTN and IP networks at the border of the communication network.
- This protocol decomposes the soft-switch into two parts for handling signaling and media transfer as listed in the following:
  - **Media Gateway Controller (MGC)** – It is a call controlling part commonly referred to as a call agent. It works as master node in the IP-PSTN

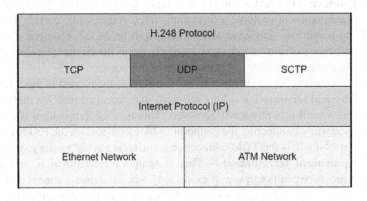

FIGURE 4.13   H.248/MEGACO protocol stack.

telephone connectivity. The main function of this segment is to establish, tear, and manage the call between two gateways in the network.

- **Media Gateway (MGW)** – This is the slave part of master/slave communication in the H.248 protocol. This part is responsible for only making media connections between two media gateways and with a few additional configurations regarding the codec, encryption, and other capabilities of the line. There is no role of this part in the establishing, managing, and tearing down of the call.

- THe H.248 protocol talks between two parts of the IP-based network – MGW and MGC – through a series of transactions of communication messages between those two entities or parts of the network managed by the protocol.

- This protocol supports a wide range of media gateways working as full-fledged media gateway or as a residential media gateway. The capabilities of a media gateway are always defined with the help of addition of a package that consists of different capabilities. The main types of packages, which offer the extension to the capabilities of the media gateway, include tone detection package, analog line supervision package, RTP package, statistic package, DTMF generation package, and many others.

- The H.248 protocol supports both binary encoding and text encoding in its messages for communication.

- This protocol describes and handles the logical identities or objects within the media gateways that are also referred to as **abstractions** in the H.248 communication model.

- The two major abstractions used in the H.248 powered communication systems within the media gateways are listed in the following:
  - **Terminations** – It is a multimedia entity that can be either source or a sink of multiple media streams. Different capabilities and parameters of the media as well as bearers are encapsulated within those terminations. The analog of termination abstractions is like ports connected to the media gateways. A termination can be created, modified, and deleted as per need. Each termination is identified with a unique identity known as TerminationID, which is assigned at the time of creation of the termination abstraction.
  - **Contexts** – Context is another abstraction used in the H.248 protocol. It is a combination of multiple terminations put together under certain data sharing conditions. Different terminations can be added, removed, modified, and moved from one context to another one through configuration commands, which will be discussed later in this topic.

- There are two types of termination abstractions as mentioned in the following:
  - **Physical termination** – This is a semi-permanent connection that remains intact until it is physically removed. This type of termination is normally used while connecting the endpoint or user available on the PSTN network. An example is the TDM connection terminating at the media gateway.
  - **Ephemeral termination** – This is temporary termination, which gets removed when not in use. It exists only when the media connection on the packet-based communication is running. An example of ephemeral termination is the real-time transfer protocol RTP stream in VoIP communication.

- The other characteristics of termination abstractions used in this protocol are listed in the following:
  - Terminations are capable of generating tones, announcements, and electrical signals.
  - They can also detect the tones, events, rings, trigger notice messages generated from the media gateway for the MGC part.
  - Media gateway creates statistical information from the termination activities, which can be accessed by MGC when needed.
- Terminations are described by two major characteristics regarding the capability set of the termination abstractions:
  - The mode of termination (send/receive or receive only)
  - The state of the termination (i.e., in-test, in-service, or out-of-service)
  - The event detection capabilities (i.e., off-hook, flash-hook, and on-hook)
  - The electrical signal generation (i.e., call-waiting tone and dial-tone)
  - The statistics capabilities (i.e., lost packets/cells, received packets/cells, and others).
- A context abstraction used in the H.248 protocol should have at least two terminations for establishing a communication for sharing the media between the two termination abstractions. The other major characteristics of a context are mentioned in the following:
  - A context uses the STAR topology to connect all terminations available in the context. This ensures the transmission of media to all participants in a context. The mixing of the streams is done on the basis of stream IDs. The same stream IDs are merged together, while the separate stream ID is separated from other streams.
  - There is one idle context, which houses all terminations that are not participating in any other context for the transmission or reception of the media streams. This context is named "Null Context".
  - When a call is set up, the terminations available in the NULL context are moved to a new context with desired capabilities, and after the completion of the call, the terminations are returned back to NULL context with default settings.
  - Media gateway itself uses the ROOT termination for communication with other gateways and MGC.
  - A context has so many capabilities to manage and control the terminations within a context. It can set different attributes for different terminations such as capabilities to receive stream, send stream, priority, connection topology, and many others.
  - A context is capable of allowing emergency calls and other such services to the terminations at any time.
  - A context is defined with a ContextID.
  - ContextID within a media gateway is unique for communication. But in different media gateways, the context can be overlapping in terms of ID numbers.
  - Wildcard context named "*" and "$" wildcard contexts are used for termination among multiple media gateways.
  - The MGC issues "$" wildcard to MGW to create a new context.

TABLE 4.4   H.248 Protocol Commands and Directions.

COMMAND	ISSUING DIRECTION
Add	From MGC to MGW
Subtract	From MGC to MGW
Modify	From MGC to MGW
Move	From MGC to MGW
Notify	From MGW to MGC
AuditCapability	From MGC to MGW
AuditValue	From MGC to MGW
ServiceChange	From MGW to MGC

- A message for communication between the media gateway and the media gateway controller is transmitted through a series of commands.
- In the H.248/MGC protocol, numerous commands are used for issuing instructions and getting a response to the requests. A few of those commands are listed in Table 4.4.

MGC = Media Gateway Controller
MGW = Media Gateway

- A group of commands is known as an "Action". The actions are used to operate a context or a set of contexts.
- The operation mode of commands in an action is in sequential order.
- Each command directly operates on the termination or a group of terminations.
- Multiple actions are grouped together to form a transaction in the MEGACO protocol.
- The transactions combined into a unit are transmitted to the corresponding entity of the MEGACO-based VoIP telephony entities.
- All actions bound within a transaction are executed in sequential order too.
- The execution of the commands stops further if any command in the sequential order fails. The following commands will not be executed due to the constraints of execution of sequential order in the system.
- In the response of transaction command, the system generates TransactionReply, which includes the information about the results of the commands in the corresponding TransactionRequest in this communication protocol.
- The unexecuted command is responded with an error code in the descriptor. The error codes are predefined and accepted by the Internet Assigned Numbers Authority (IANA). Each error code has a specified meaning for the system.

Media Gateway Control Protocol (MGCP)

Media Gateway Control Protocol precisely referred to as MGCP is the other name of H.248 or MEGACO, which has been explained in full detail in the aforementioned section. It was designed and developed for controlling the signaling and call control functions of the PSTN network to connect it to the IP network through a soft-switch. A soft-switch acts like a central office in a traditional PSTN network along with a media gateway, which integrates the PSTN and IP network in IP environment. This protocol is the successor of its predecessor protocol known as Simple Gateway Control Protocol, which was developed by the following entities:[126]

- Cisco Systems Inc.
- Telcordia Technologies (Former Bellcore Inc.)
- Internet Protocol Device Control (IPDC)

E.164 Standard

E.164 Standard is designed and developed by the ITU for the international telephone numbering scheme or plan. Prior to this standard, the numbering plan for the worldwide telephony network was based on the E.163 standard, which was superseded by the E.164 standard for the international numbering scheme.[127]

This numbering standard addresses the five basic categories of telephone numbering schemes as listed in the following:

- International numbering plan for geographical areas
- International numbering plan for networks
- International numbering plan for global services
- International numbering plan for a group of countries
- International numbering plan for the trial purposes

Three major versions of the E.164 numbering scheme standard were released for setting the standards to check the criteria, principles, and procedures for applying for the numbering schemes and other related matters in the numbering systems. Those three versions of the E.164 standards are mentioned in the following:

- **E.164.1 standard** – Recommendations for identification codes (IC)
- **E.164.2 standard** – Complete set of recommendations for country code (CC)
- **E.164.3 standard** – Comprehensive guidelines for group of countries code (GCC)
- **E.164.ARPA** – The use of international subscriber numbers into the domain name server (DNS) is done by reversing the entire number with dots. This is called telephone number mapping.

The main features and characteristics of the E.164 standard are listed here:

- The maximum number of digits used in any international telephone number should be limited to 15 digits in all services and numbering codes defined under this standard.
- The dialing numbers should be made as low as possible by the concerned administrations.

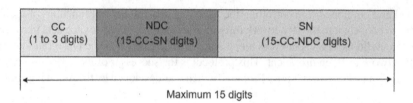

CC (1 to 3 digits)	NDC (15-CC-SN digits)	SN (15-CC-NDC digits)

Maximum 15 digits

FIGURE 4.14 International public telecommunication number for geographic area scheme.

- To identify the country's destination, the digital analysis for the telephone exchanges should be limited to a maximum of 7 digits.
- Any kinds of changes in the national numbering plans should be notified to the ITU agency at least two years prior to implementation.
- The numbering structure of the international telephone numbers will consist of three segments as shown in Figure 4.14.

CC = Country Code
NDC = National Destination Code
SN = Subscriber Number

- The separate rules and recommendations for the selection of national destination code and subscriber numbers are also defined in the standard.
- For national dialing, entering the country code is not necessary within the jurisdiction of the dialing country.
- Similarly, dialing the national code is not necessary for the same exchange local numbers.

This is also very important to note that the telephone numbers when routed through the IP network need to be translated into the addresses that the IP network can route. So, to address this issue, the ENUM standard was designed by the IETF, which maps the telephone numbers with a Domain Name System (DNS) address.

Sample Questions and Answers for Chapter 4

Q1. What is Nyquist sampling theorem?

A1. Nyquist sampling theorem is the fundamental principle of converting the analog signal into a digital signal for the pulse code modulation in telecommunication technology. The Nyquist theorem states that:

"Any analog signal can be converted into a digital signal through pulse code modulation by taking the discrete samples at equal interval either twice or higher times the maximum frequency of the analog signals. The signal taken as much as twice (min) the frequency of analog signal can be reconstructed into the analog signals without any loss of the information in the analog signal. This theorem also says that the greater number of samples would result in a better quality of the signal conversion".

Q2. What is *companding* in telecommunication?

A2. In telecommunication and signal processing, *companding* is a method of mitigating the detrimental effects of a channel with a limited dynamic range. The name *companding* is derived from the processes of compression and expansion of the voice signal.

Q3. Name the two major signal compressing Laws that are extensively used in telephony systems worldwide.

A3. There are two major laws extensively used in telephony systems worldwide:
- μ-Law Companding
- A-Law Companding

Q4. Name the three major nodes that the SS7 signaling system is consisted of.

A4. An SS7 signaling system consists of three major nodes in the entire signaling system. They are listed in the following:
- Service Switching Point (SSP)
- Signaling Transfer Point (STP)
- Service Control Point (SCP)

Q5. What are the main services offered through the ISDN network?

A5. The main services offered through the ISDN network include the following:
- Voice services
- Data services
- Text services
- Image services
- Video conferencing services
- Audio conferencing services

Important Data Communication Protocols

5

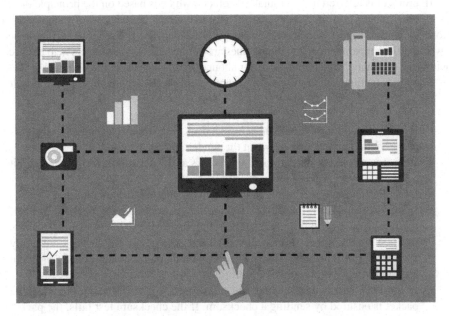

Data communication is a modern domain of communication in which the digital data is transmitted between a source and a destination or a sender and a receiver through different types of media such as copper wire, fiber optics, computer buses, electronic circuits, storage devices, and wireless channels. The entire data communication – point-to-point or point-to-multipoint – is governed by different communication protocols. Generally speaking, those protocols are known as data communication protocols.

In this chapter, we will focus on the different types of modern data communication protocols that work either at network and upper layers (application-specific) protocols such as routing protocols, transport protocols, and other application-specific protocols that are used by the modern data services in the environment of the latest Internet-based technologies.

DOI: 10.1201/9781003300908-7

145

INTERNET PROTOCOL (IP)

IP is the most powerful and the most popular protocol in modern data communication based on IP addresses. In other words, we can say that the IP is the backbone of our modern communication for sending and receiving information over the Internet and private networks (based on IP protocol).

IP protocol is that governs the transmission and reception of data packets from one point to another point at the network layer of the OSI model of communication. This protocol uses the IP addresses as the core addressing scheme to encapsulate the datagram into a packet and route the packet to the destination. This entire function is performed by adding the IP header to the packet, which contains different parameters required for the transmission and reception of the packet from one point to another one over the network.

IP protocol is referred to as a routable protocol, which is based on the principle of "*The best effort protocol*". At the sender end, IP protocol wraps the payload of the message or information into a header, which is known as the IP header. The IP header has different fields in which different parameters and their values are defined. Those parameters help the packet route to the destination and tell the receiver about different aspects of the wrapped packet.[128]

The main features, capabilities, and functions of the IP are mentioned in the following list:

- It is a connectionless protocol because it does not establish an end-to-end physical or virtual connection between the sender and the receiver.
- IP protocol is a routable protocol that works at the network layer. We can call it a network layer protocol too.
- It handles addressing, routing, and data encapsulation for transmission.
- It does not support reliability by tracing the lost or damaged packets and correcting them or resending them. The reliability of the delivery of the packet is handled by the upper layer (transport layer) protocols, which are commonly used with the IP protocol. The most common name of the upper layer protocol used with the IP protocol is TCP protocol.
- In the IPv4 addressing scheme, the correctness of the IP header added to the packet is ensured by sending a checksum. If the checksum test fails, the packet is discarded by the routing nodes in the network. But, in IPv6, the use of checksum is also not available and the reliability fully depends on the lower-level layer protocols at the data link layer and below.
- A packet used by the IP protocol is divided into two parts – IP header and packet payload. The payload is the data and control information received from the upper layers, and the IP header contains the information about many other parameters that are used for the delivery of the packet over the network in an IP network. Figure 5.1 shows the packet header fields.
- The IP header for IPv4/IPv6 version consists of the following fields of information:[129]
  - **Source IP address** – This field is a 32-bit field for IPv4 and 128-bit for IPv6. This field contains the IP address of the sender node.
  - **Destination IP address** – This field is of the same length of the source IP address field. This contains the IP address of the receiver node.

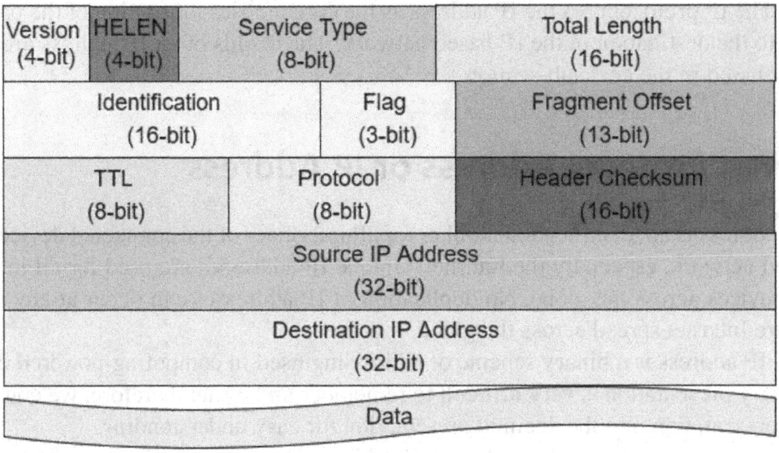

Version (4-bit)	HELEN (4-bit)	Service Type (8-bit)	Total Length (16-bit)
Identification (16-bit)		Flag (3-bit)	Fragment Offset (13-bit)
TTL (8-bit)		Protocol (8-bit)	Header Checksum (16-bit)
Source IP Address (32-bit)			
Destination IP Address (32-bit)			
Data			

FIGURE 5.1   Packet header fields.

- **Protocol** – This field contains the information about the protocol that a receiving end should pass the packet. For example, UDP, TCP, or others.
- **Identification** – The Identification field is a 16-bit field. It is used to identify the fragments within a packet so that the data can be reconstituted properly.
- **Version** – It is a four-bit field to tell the receiver about the version of IP used in the packet header.
- **Internet header length** – This is a 4-bit field, which is also referred to as HELEN and IHL. This field shows how many bytes are there in the header.
- **Type of service** – This is a 3-bit field used for indicating the type of service such as time-sensitive or any other type of service. This field is also referred to as Differentiated Service Code Point or DSCP.
- **Total length** – This field range shows the total dimension of the packet including the header of the packet. The size of the total length of the packet can range between 20 and 65,535 bytes.
- **IP flag** – This field is used to identify and control the fragments in the packet. It is a three-bit field.
- **Fragment offset** – This field is used to show the number of data bytes before any fragments in any datagram encapsulated in the packet. The value of this field ranges between 8 and 65,528 bytes.
- **Time to live** – It is also referred to as the TTL field. The main purpose of this field is to set the living time of the datagram before it is discarded. The range of this field is from 0 to 255.
- **Header checksum** – It is a 16-bit field, which is used for checking any error in the header of the packet. If the value of the checksum does not match the header value, the packet is discarded.
- **IP options** – This field is used in IPv4 packets. It indicates the value of timestamp, security, record route, and others.
- The transmission of a packet is routed through different routing protocols based on IP addressing schemes.

- The IP protocol uses the IP address as the core metrics for routing of the packet to the destination in the IP-based network. The details of an IP address are discussed in the next sub-section.

Internet Protocol Address or IP Address

The IP address is an identification number for all interfaces of the connected devices on an IP-based network, especially the Internet. Unique IP addresses are used for all interfaces of the devices across the globe. No duplication of IP addresses can occur at any time on the entire Internet spread across the globe.

An IP address is a binary scheme of addressing used in computing-powered devices. The binary presentation is very difficult to remember and write; therefore, we convert the binary presentation into the decimal presentation for easy understanding.

There are two types of IP addresses used in modern communication systems based on the IP protocol for routing the packets and messages across the networks or network of the networks referred to as the Internet. Those two types of Internet addressing schemes are listed in the following:

- IP addressing scheme version 4 (IPv4)
- IP addressing scheme version 6 (IPv6)

IP addressing scheme version 4 is a 32-bit binary long addressing scheme. This 32-bit long binary presentation is divided into two parts as listed in the following:

- Network prefix
- Host number or host address

IPv4 Addressing Scheme

IPv4 is a 32-bit addressing scheme. The total number of IP addresses of the IPv4 scheme is 2^{32}, which generates as many as 4,294,967,296 unique IP addresses to be used for the connected interfaces of the computing-powered devices. Thus, you can connect about 4.2 billion devices or interfaces simultaneously. The bigger IP-based network is the Internet, which is expanding very fast – so fast that this number of IP addresses looks very small to cater to the increasing need of the IP addresses. This is the reason that the new version of IP addresses known as IPv6 has been introduced and implemented in the present communication systems along with the existing IPv4 addresses.

The main features of IPv4 are mentioned in the following list.

- 32-bit IPv4 address is presented as four octets of binary digits as shown in the following:

  101010110: 101010110: 00001010: 10000110

- Writing as well as remembering the aforementioned presentation of IPv4 is very difficult. So, it is written in decimal form by converting the value of binary octet into decimal and separating the value of each octet by a dot (.) presentation as shown in the following:

125.38.100.209

- The value of each octet will range between 0 and 255.
- An IPv4 address consists of two parts – network address and host address. Both numbers vary with respect to the class of the IP address.
- To differentiate the IP networks, the sub-net masking is used, which is a mask to hide the network ID so that the hosts can be identified and separated.
- The scheme of sub-net masking is also presented in the binary digits and decimal digits. The format of a sub-net mask is shown in the following:

255.255.255.000

- IPv4 addresses are classified into the following five classes:
  - Class A
  - Class B
  - Class C
  - Class D
  - Class E
- Every class of IPv4 addressing scheme is defined and classified according to different functions, purposes, number of networks, number of hosts, and other parameters.
- The details of those parameters of all classes of IPv4 addressing systems are mentioned in Table 5.1.
- **Class A** – Class A is the first group of IP addresses. It has 8 bits specified for the network addresses, and the remaining 24 bits out of the total 32 bits are specified for the host addresses. The number of network addresses that Class A can accommodate is 128 because $128 = 2^7$. The total number of host IP addresses in a single network would be equal to 2^{24} or 16,777,216. Thus, the total number of IP addresses that can be accommodated in Class A would be 2,147,483,648. Class A is the class of IP addresses that can accommodate most of all other classes. One series of IP addresses (127.0.0.0) in this class is reserved for the loopback addresses.
- **Class B** – This is the second-largest IP group, which can accommodate as many as 1,073,741,824 hosts. It can accommodate as many as 16,384 networks with a capacity of 65,536 hosts per network.
- **Class C** – This class has the capacity to accommodate 536,870,912 hosts in total. It can house as many as 2,097,152 networks, and each network can accommodate as many as 256 hosts. The capacity of the networks belonging to class C is very low and suitable for small networks.
- **Class D** – It is not a class specified for normal communication purposes. It is specified for only multicasting applications for propagating the information within a group of users. This IP cannot be used as hosts. The device configured with this IP address is used for multicasting the information. The Class D IP range is 224.0.0.0 through 239.255.255.255.
- **Class E** – This class was not specified and was reserved for future use and scientific research. The use of this IP on the public network is prohibited. One

TABLE 5.1 Classes of IP Addresses.

CLASS	START IP ADDRESS	END IP ADDRESS	NETWORK ADDRESS BITS	HOST ADDRESS BITS	NUMBER OF NETWORKS	NUMBER OF HOSTS PER NETWORK	TOTAL NUMBER OF HOSTS IN ENTIRE CLASS
Class A	0.0.0.0	127.255.255.255	8	24	128	16,777,216	2,147,483,648
Class B	128.0.0.0	191.255.255.255	16	16	16,384	65,536	1,073,741,824
Class C	192.0.0.0	223.255.255.255	24	8	2,097,152	256	536,870,912
Class D (multicast)	224.0.0.0	239.255.255.255	Unspecified	Unspecified	Unspecified	Unspecified	268,435,456
Class E (reserved)	240.0.0.0	255.255.255.255	Unspecified	Unspecified	Unspecified	Unspecified	268,435,456

IP 255.255.255.255 is used for the broadcasting purpose. The rest of the IPs in this class are reserved for special purposes. The range of IPs of this class is 240.0.0.0 through 255.255.255.255.

- **Private IP Addresses** – A certain number of IP addresses from these classes have been reserved for private use. Those IPs are known as private IP addresses. Those private IP addresses have been taken from multiple classes. The ranges of the private IPs are listed in the following:
  - Class A – From 10.0.0.1 to 10.255.255.255
  - Class B – From 172.16.0.0 to 172.32.255.255
  - Class C – From 192.168.0.0 to 192.168.255.255

IPv6 Addressing Scheme

With the explosive growth of the Internet, experts in the industry started projecting the insufficiency of the number of IPv4 addresses to cater to the increasing demand for IP addresses in the future. IPv6 is a 128-bit format of IP addresses, which is much bigger than IPv4, which uses 32 bits for defining the unique IP of a host and networks.[130]

IPv6 can accommodate 340 decillion IP addresses, which is a huge number of IP addresses. One decillion is 10^{36}. Theoretically, it will not run out in many centuries to come as far as the existing technology paradigm prevails. The format of IPv6 addressing notation in hexadecimal format is shown in the following:

1BAB : FEFF : 0100 : 3EAE : 01BA : 00FF : DB72 : 2B2A

The binary presentation of an IPv6 address based on 8 blocks of 16 binary digits is shown in the following:

0101010110101011 : 0101010110101011 : 0101010110101011 : 0101010110101011 : 0101010110101011 : 0101010110101011 : 0101010110101011 : 0101010110101011

The main features and characteristics of the IPv6 addressing scheme are mentioned in the following list to provide you with a better overview.

- It is a 128-bit long addressing scheme
- All 128 bits are distributed into 8 blocks of 16 binary digits
- Each block of IPv6 address is separated by a colon sign (:)
- The binary notation is difficult to write. So, the formal presentation of IPv6 is done in hexadecimal notation.
- The hexadecimal 4 digits can accommodate 16 digits of binary digits. So, the 8 blocks of 4 digits of hexadecimal notation present the entire address.
- Each block is separated by the same colon notation.
- IPv6 can accommodate as many as 3.4×10^{38} unique interfaces on the networks.
- All consecutive zeros in any IPv4 address can be replaced with double colon marks (::). This notation cannot be used more than once in any presentation.

- The loopback address of IPv6 in hexadecimal notation is given in the following:

0000:0000:0000: 0000:0000:0000: 0000:0001

- The aforementioned presentation of loopback address can be shortened as:

:: 0001 or::1

- The private IP addresses in IPv6 are known as Unique Local Addresses (ULA).
- Like IPv4, the private IP addresses in IPv6 cannot be used on the public network. The bunch of private IP addresses is given in the following:

FC00:: /7

- IPv6 uses the Neighbor Discovery Protocol (NDP) for autoconfiguration of IPv6 addresses and advertisement of the network prefix.
- The IP header of the IPv6 address is just double the IPv4 address.

IPv6 addresses can cater many types of existing services such as the IoT, which is expanding exponentially and needs a huge number of unique IP addresses for almost everything in the modern world. Any kind of future needs of IP addresses can also be catered easily with the help of this gigantic pool of IP addresses.

Transmission Control Protocol (TCP)

TCP is a transport layer protocol that communicates with the applications working on the application layers and facilitates network layer protocol – mostly IP protocol, to transmit the data in packets without error correction and other capabilities. The TCP protocol is a bridge between the communication between applications and networks.

TCP protocol can act as a communication protocol within a private intranet as well as extranet systems without public routing. The TCP uses port numbers for different types of applications to communicate with that particular application at the upper OSI layers. On the other hand, it controls the segments, error control, flow control, and other functions while transmitting the datagrams over the IP protocol in the form of packets.

TCP protocol was introduced in 1974 through a paper. The name of the TCP protocol was "*Protocol for Packet Network Communication*" at that time. This is a very important protocol that works with the IP protocol in the modern Internet. The use of TCP with IP protocol is so prominent that the name of both protocols is used simultaneously like TCP/IP protocol stack.

TCP protocol takes the message from the upper layer applications such as mail server, ftp server, or any other. It breaks the messages into small segments for efficient transmission over the internet. Each segment is numbered with a sequence number so that the reassembling of the segments can be done easily. The error correction checksum makes sure that the packet transmitted and received is not corrupted or damaged. It can perform many other functions and activities as described in the TCP header shown in Figure 5.2.

The explanation of the fields in a TCP header will be described in the main features and characteristics list. The main features and characteristics of TCP are mentioned in the following list:[131]

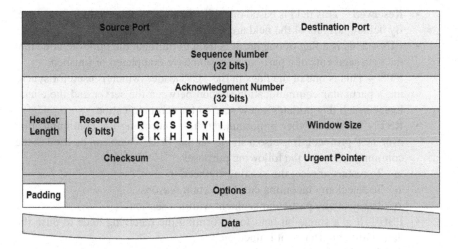

FIGURE 5.2   TCP header fields.

- This protocol belongs to the transport layer of the OSI model of communication.
- It is a connection-oriented protocol for reliable communication of data packets over the unreliable IP, which does not guarantee the reliable delivery of the data packet to its destination.
- The sequence is used to rearrange the segments of the packets broken for easy transmission over the internet through different routes.
- The connection between two endpoint clients for connection-oriented transmission is established through port number in the TCP packet.
- The TCP protocol supports end-to-end communication.
- It supports error detection and correction mechanism through retransmission of packet segments.
- It is based on the server-client communication mechanism in full-duplex communication mode in which it can perform the role of server and client simultaneously.
- The minimum length of a TCP header is 20 bytes and the maximum length is 60 bytes.
- The fields of a TCP header are described in the following:[132]
    - **Source port** – The source port is defined at the sending device. It is a 16-bit field.
    - **Destination port** – The destination port is also a 16-bit field, which defines the port number of the receiving device.
    - **Sequence number** – The sequence number is a 32-bit address, which shows the sequence of the data segments in a particular communication session.
    - **Acknowledgment number** – This field is also a 32-bit field, which denotes the sequence number of the next segment in a particular session. This expected number is also used as an acknowledgment in the communication that the previous bit has reached the destination in sound condition.
    - **Header length** – This is a field that denotes the size of the header length. It is also known as the data offset field. It is a 4-bit field.

- **Reserved** – This field is based on 6 bits. It has been reserved for future use. By default, all bits of the field are zeros.
- **FIN** – This is a flag of one bit. It is used to intimate the opposite node that the data segments of a particular session have completed or finished.
- **SYN** – This is one of the flags in the TCP header, which is used for synching a particular communication session between the server and the client. Every synch flag initiates a new session for communication.
- **RST** – This is another important flag for controlling the communication over TCP protocol. It is also a single-bit flag field, which can be used in the communication for the following purposes:
  - o To restart or reset the existing connection
  - o To reject any incoming call for certain reasons
  - o To reject any particular segment within a session
- **PSH** – It is a single-bit field for informing the receiving node to push the data without buffering it immediately.
- **ACK** – This single-bit flag is used for notifying the receiver node that the mechanism of acknowledgment in the TCP header is set for sending an acknowledgment. If the value of this flag is set to zero, which means that the TCP header does not want the acknowledgment of the sequence number.
- **URG** – The urgent flag is also a 1-bit field. It is used for indicating the importance of the data in the **Urgent Pointer** field of the TCP header. The value 1 in the field indicates that the urgent pointer data is important and should be processed on a priority basis. If the value of this flag is zero, then no urgency of the data in the urgent pointer field to process immediately.
- **Window size** – It is a 16-bit field, which is used for the control of the flow to maintain the smoothness and control the congestion of data in the communication channel. This window indicates the status of the allocation of buffer at the receiving-end node. This information provides the status of the receiving end's node capacity to be able to handle the data properly.
- **Checksum** – The checksum field is a 16-bit field in the TCP header. It is used for error detection and correction. Any error is detected through an algorithm to calculate the total bits of the segment.
- **Urgent Pointer** – It is a 16-bit field in the TCP header. It shows the amount of urgent data in bytes when the URG flag is set to 1 in the flags field.
- **Padding** – This field is used for denoting the receiving-end that the TCP header has ended and the data of the packet has started on a 32-bit boundary. It is all zeros field in the TCP header.
- **Options** – This field is a 32-bit field. It is used for indicating different purposes. The main information about the options conveyed by this field includes the following:
  - o End of option list – Known as "Kind 0", which means the field will be 00000000.
  - o No operation – Known as "Kind 1", which means the field will be 00000001.
  - o Max segment size – Known as "Kind 2", which means the field will be 00000010 in the first octet, and the size will be indicated in the next octet.

- The addressing of the TCP end-nodes is done through the **Transport Service Access Point (TSAP)**. It is a long series of port numbers ranging from 0 to 65,535.
- Those TSAP ports are classified into three major categories as listed in the following:
    - **System TSAP ports** – From 0 to 1,023 port number
    - **User TSAP ports** – From 1,024 to 49,151 port number
    - **Dynamic ports or Private TSAPs** – From 49,152 to 65,535 number.
- The establishment of a connection over the TCP protocol is done by a client, which initiates the connection by sending a segment with a sequence number for the session. The server receives the sequence number and acknowledges it with its own sequence number for the confirmation that the communication session is ready to be established. The client acknowledges the server's segment by an acknowledgment message. The communication session has been established between the two points.
- The release of the end-to-end communication session is accomplished by sending a TCP segment with the FIN flag field set to 1 by either sender or receiver. The receiver of the segment with a flag will also send the acknowledgment segment with the FIN flag set to 1 in the response. The communication session ends at both ends.
- The TCP protocol is also capable of using the feature of multiplexing, in which multiple ports are multiplexed to receive data from different ports like Telnet, HTTP, FTP, SMTP, and others. All these ports can communicate on a single end-to-end virtual connection established by the TCP protocol through different sequences of segments in different sessions.
- The TCP protocol uses different types of timers for time management of TCP and session communication. A few of those timers are listed in the following:
    - **Timed-Wait Timer** – Used for waiting the end of the communication
    - **Persist Timer** – It is used to maintain the connection for a certain time when the windows size is sent as "0" to control flow. The persist timer will wait for a certain time before receiving the segment with window size other than "0".
    - **Keep-Alive Timer** – This timer is used for checking the validity and integrity of the connection in TCP communication.
    - **Re-Transmission Timer** – This timer is used to wait for a certain period before retransmission of the same packet, if the acknowledgment of the data sent earlier through a segment does not reach.
- The TCP protocol used three major algorithms for controlling the flow and congestion. Those algorithms are listed in the following:
    - Slow start algorithm
    - Timeout React algorithm
    - Additive increase and multiplicative decrease algorithm.
- All of the aforementioned congestion control algorithms are governed by the window size field by sending the information of windows size to the sending device.

User Datagram Protocol (UDP)

User Datagram Protocol, commonly known as UDP, is another very important protocol that deals with the communication at the transport layer of the OSI model of communication. It is a connectionless protocol, which does not establish an end-to-end connection between the receiver and the user. This is the reason that it is called an *unreliable* transport protocol. This protocol is extensively used in real-time applications that are time-sensitive. The use of a reliable protocol based on retransmission of data packets after long delays are meaningless for time-sensitive applications such as real-time voice call, video call, and live streaming applications. In those applications, the use of UDP is the most suited protocol, which does not retransmit a lost packet (and, the speed is the main concern). The output of the real-time and time-sensitive applications will look a bit better with a lost packet than the output that has a packet received after a long delay. It will even make the output unclear.

The use of unreliable transport protocols like UDP also increases the performance and quality of the data communication due to the reduced overload on the communication link to check and acknowledge each packet as it is done in the TCP protocol. So, this protocol is highly suitable for increasing the performance of the communication with its unreliable nature.

The UDP protocol header is very smaller than the TCP protocol header due to the absence of many error correction and acknowledgment fields in the UDP header. This header does not have fields for different types of tags to inform the receivers of certain actions and other activities. The UDP header is an unreliable protocol, so it does not support numerous other fields such as urgent pointer and others, which are associated with the maintenance of QoS or defining the priority data processing functions.[133]

The main fields of the UDP protocol header are shown in Figure 5.3.

The UDP protocol header is very simple and short. It consists of four fields and just 8 bytes of data. The TCP header consists of 20 bytes minimum and maximum data of 60 bytes. Thus, the header itself is a substantial burden on the bandwidth of the communication in the case of TCP communication protocol. All the four fields of UDP protocols are listed in the following:

- **Source port** – This field consists of 2 bytes of data. This field takes the source port number of the application communicating with the transport layer UDP protocol.

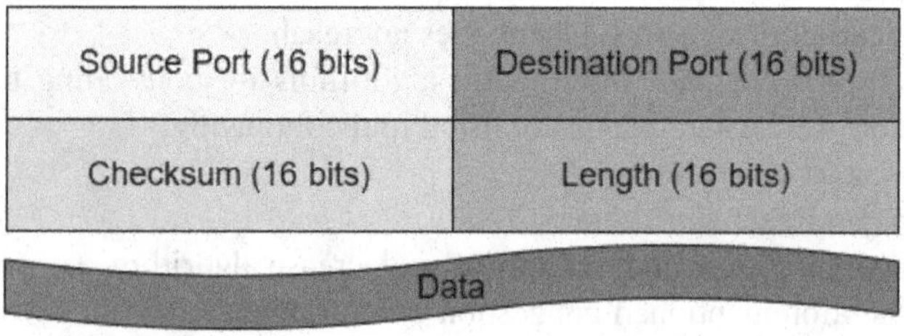

Source Port (16 bits)	Destination Port (16 bits)
Checksum (16 bits)	Length (16 bits)
Data	

FIGURE 5.3 UDP protocol header.

- **Destination port** – This is the port number of the receiver application or machine to which the packet is destined. This field is also a 2-byte field.
- **Checksum** – This field contains the value, which is also known as the checksum value generated by the sender before the packet is sent out. It is used for checking the integrity of the data packet in the UDP. This field is optional in the IPv4. In that case, all of the bits of this field are set to zero value.
- **Length** – This field denotes the length of the entire UDP data packet. The length includes the data attached to the UDP packet and header of the UDP protocol. This field consists of 16 bits or 2 bytes. The minimum value of a UDP packet is set to 8 bytes, which is the total length of a UDP header without any data.

The most common web-based modern applications or services that use the UDP as its transport layer protocol include the following:[134]

- Routing information protocol (RIP)
- SNMP
- Domain Name Service (DNS)
- Kerberos Authentication Protocol
- Trivial File Transfer Protocol (TFTP)
- Network Time Protocol (NTP)
- Network News Protocol
- Dynamic Host Configuration Protocol
- Bootstrap Protocol
- Real-Time Streaming Protocol (RTSP)

The UDP protocol is very popular in many services, especially multimedia services that are time-sensitive and real-time applications. The other main features and characteristics of the UDP communication protocol are listed in the following:

- This protocol is known as a stateless protocol.
- Checksum is an optional field in the UDP header. When used without checksum, it is known as a Null protocol too.
- Awesome for video streaming, video gaming, VoIP, and video conferencing services.
- The orderly delivery of the UDP packets is not guaranteed by this protocol.
- It is a connectionless protocol at the network layer.
- It does not support the feature to control the congestion in the bandwidth or routes.
- Considered the most suitable protocol for one-way data flowing applications.
- No capability or feature to control the QoS, acknowledgment, and error detection and correction at all.
- It saves substantial bandwidth as compared to its other counterpart (TCP protocol).
- For error reporting, the UDP is fully dependent on the ICMP and IP protocols.
- A great protocol for multicasting services.

- Different types of application-level functions like tracing a route, time stamping, route recording, and others use the UDP protocol.
- For data transmission, UDP does not need any kind of three-way handshaking because it does not support end-to-end connectivity features.
- Protection against the duplicate packet is also not supported by this protocol.
- The UDP packet is not so secure due to the absence of many control functions for the delivery and acceptance of the packets. This is the reason that many hackers exploit this vulnerability of UDP for unleashing distributed denial of service (DDoS) attacks.

TOP APPLICATION-SPECIFIC DATA COMMUNICATION PROTOCOLS

With the advent of modern web technologies powered by the Internet and other modes of communications through the latest communication technologies, the number of application-level protocols has increased substantially. Numerous types of applications have been developed that use different kinds of application-specific protocols for communicating with the lower layer protocols such as transport, network, and other layer protocols.

The major applications that use different types of protocols for establishing communication with the lower layers of the OSI model of communication include:

- Websites and web applications
- Email services
- File sharing and transferring services
- Remote device access services

All of the aforementioned services and many others use different types of application-specific communication protocols. A few of those important protocols are mentioned in the following.

Hypertext Transfer Protocol (HTTP)

Hypertext Transfer Protocol, most commonly known as HTTP protocol, is an application-specific protocol that works on the WWW communication system. This communication system is based on the client and server communication model. Thus, HTTP is a client/server protocol based on the two types of messages as listed in the following:

- HTTP Requests
- HTTP Responses

It is a very simple and easy-to-understand protocol in plain text formats, which can be easily understood by the readers with a little knowledge of basic HyperText Markup

Language (HTML). But, with the adoption of encryption in the second version of HTTP 2.0, the use of plain text in the file has been replaced with the binary format of data.[135]

In HTTP communication, the clients are normally the web browsers installed on different computers powered by different operating systems and platforms. A few top HTTP client software – browsers – are listed in the following:

- Mozilla Firefox
- Google Chrome
- Microsoft Edge
- Microsoft Explorer
- Apple Safari

In certain and very special cases, the HTTP clients can be other than web browsers such as search engine indexing crawlers, software debuggers, and other such special-purpose tools that work on the basis of HTTP request/response communication mechanism.

The HTTP protocol is the foundation of the entire communication on a worldwide web system. The HTTP request initiates the communication through any supported browser to request the server to send a resource file, which is normally an HTML file. This file is sent by the server in the response to the request initiated by the client browser. The file is opened and parsed to send more requests for additional resources required to complete the loading of the web resources. In these requests, the scripts are run to request more files such as images, videos, audios, style sheet, layout, and other items. The clickable links are hypertexts, which can be triggered by clicking them to initiate another HTTP request for a certain resource to send from the server.

Any modern web browser has an inbuilt software object based on JavaScript referred to as JavaScript object. This object is named **XML HTTP Request Object (XHR)**. The purpose of this JavaScript object is to initiate a request for some resources and collect the data received back from a server in the response to the request. The main characteristics of XHR are listed here:

- An object, which can be used in HTTP clients as well as with many other protocols.
- Capable of processing numerous types of data such as Cascading Style Sheets, HTML, JavaScript Object Notation (JSON), Extensible Markup Language (XML), and plain text.
- It is the underlying concept for JSON, and AJAX for webpage updating.
- Updates the page without reloading the entire web page.
- It can request for data and receive data after the page has been loaded.

It can communicate with the server in the background by sending data.

In this entire communication of request and response between the client browser and the server, there may be one or more servers located on one single or multiple machines for different resources. For example:

- **Web server** – for html webpage resources
- **Video server** – for sending video files

- **Photo server** – for sending images

The complete path of HTTP communication involves three basic entities and numerous (variable) supportive entities. The basic entities in this communication include the following:

- Resource servers
- Browser clients
- Internet cloud

As the auxiliary entities to accomplish the communication over the www network, we need multiple types of proxies, which may be servers or any other entities. The main proxy entities in HTTP communication may include the following:

- Multiple routers
- Multiple switches
- Multiple modems
- Other nodes

All those proxies do not play any active role in protocol level processing of the data in the requests and responses, but they act as gateways for transporting at the transport and network layer. They can also play the role of a catching proxy for facilitating communication with better speed and performance in the entire process.

The browsers (clients of HTTP-based communication) can keep executing the scripts on the requested resources (web pages), and new requests are generated for the further sub-resources to update the page accordingly. Any dynamic content at the browser is automatically updated by running the scripts on the browser. Numerous latest technologies based on web browsers have been introduced to update the changes in the websites without any reloads of the entire pages. The example of such the latest model is also introduced with the help of modern JavaScript platforms such as React JavaScript Library, which uses components and contained execution of the code to update the files on the websites without any reloads of the entire pages.

The HTTP request and response have different parameters and fields, which perform their respective activities. Let's explore them with the help of diagrams of the headers of HTTP requests and responses used in this communication.

The format and the components of an HTTP Request originated from a browser client are shown in Figure 5.4.

An HTTP request consists of four major parts as shown in Figure 5.4. Those parts are explained here with full details:[136]

- **Method** – HTTP uses different types of methods that are defined for a particular action to perform in the HTTP communication. These methods are already named and defined for a certain activity. Those methods are also referred to as HTTP verbs but not very frequently. A few very important methods are listed in the following:

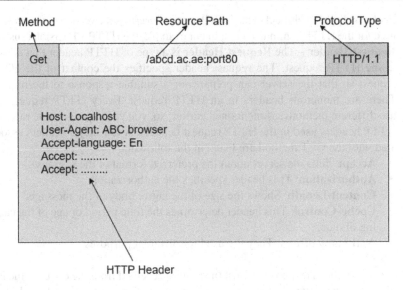

FIGURE 5.4   HTTP request format and components.

- **GET** – This method is used to retrieve data in the request. In this method, the representation of a specific resource is done in the method as shown in the aforementioned figure.
- **POST** – It submits the entity to a specific resource. This method is normally used for changes in state on the server.
- **HEAD** – This is similar to the GET method, but it does not have a response body.
- **PATCH** – It is used for partial modification of the resources.
- **TRACE** – It is used for a loop-back test along the path to the resource.
- **OPTIONS** – This method is used to specify the types of communication paths to the specified resource.
- Others – A few other methods are also used like DELETE, CONNET, and PUT.
- **Resource Path** – The resource path field specifies the location of the target resources. The target resource may be an HTML file, image, video, or any other supported content for HTTP communication. All resources used in the HTTP communication are identified by a unique ID referred to as **Uniform Resource Identifier** (**URI**). There are two major types of URIs, which are used for describing any web resource through a location or name. The syntax of URI differs with their respective types. Two types of URIs are listed in the following:
  - **Uniform Resource Locator** (**URL**) – This is a unique name, which is also known as web address that we type into the browser's address bar.
  - **Uniform Resource Names** (**URN**) – This type of URI specifies the resource with a particular name in a specific namespace or domain.

- **Protocol Type** – This field in the HTTP request describes the version of the protocol used for the HTTP communication. In our example, the HTTP 1.1 version is used.
- **Request Header** – The **Request Header** is a type of HTTP header that is used in the HTTP request. The request header specifies the context of the HTTP request so that the server can prepare for a suitable response to the request. There are numerous headers in an HTTP request. Every HTTP request can use different methods, as mentioned earlier; so, you can expect a wide range of HTTP headers used in the HTTP request headers. A few major headers associated with the GET method are listed in the following:[137]
  - **Accept**: Tells the server about the preferred format of the response.
  - **Authorization**: This header specifies the authorization.
  - **Content-Length**: Shows the size of the entire body of the message.
  - **Cache-Control**: This header determines the time period or age of the caching of data.
  - **And many others**: They are used for different purposes.

The server tailors a response against the request received from the client in the HTTP communication. The tailored response is sent in accordance with the preferred formats, and other factors provided in the shape of HTTP headers in request. The typical response generated and sent by a server is shown in the schematic diagram shown in Figure 5.5.

Like the HTTP request, the HTTP response from the server to the client also consists of the four major parts, which are explained here:

- **Protocol** – The protocol section of the HTTP response message describes the version of the protocol used in the response. In our case, it is the HTTP 1.1 version of the protocol as shown in Figure 5.5.

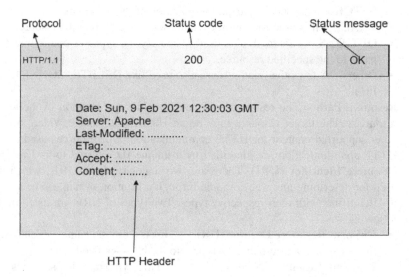

FIGURE 5.5 HTTP response format.

- **Status Code** – The status code field specifies whether the requested operation was successful or not by providing a specific code number. Every code number used in the HTTP communication has a specific meaning with full reasons. In our case, the operation against the GET request is successful because it was notified through a status code "200", which shows that the operation is successful. Another very popular code for unsuccessful operation is 404 code. This code means that the page requested in the http request was not found; so, the operation failed.
- **Status Message** – This message is related to a short explanation or description of the code, which indicates the status of the operation.
- **Response Header** – The response header is a set of HTTP headers, which provides the detailed information about the response tailored in accordance with the request sent through GET or any other method in the http request. There are numerous HTTP headers, which are generated in response to a particular request sent through a particular method. The first line of the response message is known as the Status Line, which shows protocol version, failure reason, and other factors. The response headers consist of different types of HTTP headers. The typical HTTP headers of a response generated by the server against the HTTP response based on the GET method are shown in the following list:
  - **Accept-Ranges** – Directs the client to start the operation from where it was left rather than re-downloading the data from the beginning.
  - **Age** – This is the time since the response was generated at the server.
  - **ETag** – This indicates entity tags. It is normally used for comparison with other entities.
  - **Content-Length** – It describes the size of the entity body.
  - **Connection** – Instructs the proxies whether to share a connection or not with other entities in the communication.
  - **Content-Encoding** – This header provides the information about the encoding of the content.
  - **Accept-Language** – This header is used to inform about the language of the content.
  - **And many others** – A large number of other headers are also used for providing different kinds of information and instructions.

Types of HTTP Headers

The headers play a very vital role in the HTTP protocols for establishing highly managed, secure, and controlled communication between the server and the client. The HTTP headers are used for sharing additional information between the client, the server, and proxies that take part in the communication over the HTTP communication. The structure of the header consists of two parts – name and colon sign (:). The name is case sensitive in HTTP protocol headers.

The HTTP protocol headers can be classified into the following major categories, which have a large number of headers within every category.

- **General headers** – These headers can be used in both the requests and responses.
- **Entity headers** – These headers provide the information about the body of the content.
- **Request headers** – The detailed information about the source and about the resources that the client is requesting for fetching is provided by these headers.
- **Response headers** – These headers are commonly used for providing the details about the server and its location and details about the resources provided by the server against the request originated from the http client.

Similarly, the headers of the HTTP protocol can also be classified in terms of the handling of the proxies in the communication. A few such categories are listed here:

- End-to-end HTTP headers
- Hop-by-hop HTTP headers

On the basis of the aforementioned categories of the headers, a large number of http headers have been defined by the HTTP protocol for performing specific tasks in the communication. As said earlier, the HTTP protocol is extensible and evolving; so, more headers and other features can be added in the future accordingly.

Main Features of HTTP Protocol

- HTTP protocol was designed in the early 1990s for web communication.
- HTTP is a simple and human-readable protocol. The advanced versions are based on binary expression, which is not human readable. The HTTP/2 encrypts the messages in binary expression, which is not human readable.
- It is an extensible protocol, which means that it can fetch different types of data such as video, scripts, data, hypertext documents, and many others. New capabilities can also be introduced in the HTTP protocol by using its nature of extensibility.
- It is also capable of fetching small parts and piece of documents to update the web pages as per requirements or on-demand.
- Entire protocol consists of three major parts – client, server, and proxies.
- The clients can be web browsers, debugging software, crawlers, and others.
- Over the time, this protocol has evolved. Now, a few server-initiated messages can also be used for requests, which were initiated by the clients only in the past.
- Servers may be virtually a single entity. But the combination of multiple servers is called a server in HTTP-based communication.
- Proxies can be other devices acting as gateways and transmitters. There are two types of proxies – transparent and non-transparent. The transparent proxies forward the message without any alteration in the messages. And the non-transparent proxies may make some changes in the message and then forward it to the next proxy or server.

- The main functions of a proxy may include the following:
- Load-balancing by allowing different servers to respond.
  - Logging events on a storage
  - Establishing control over different types of resources through an authentication process
  - Caching the resources in public or private domains
  - Filtering activity to control viruses and establish parent control
- In the HTTP protocol, there is no connection between two requests; therefore, it is called a stateless protocol. On the other hand, certain cookies are used for establishing a *stateful* session of communication; thus, it is also known as connection-orient with the help of cookies.
- The connection orientation feature is handled by the network layer protocols like TCP protocol. It also uses the UDP protocol for certain applications.
- By default, HTTP/1.0 opens a separate TCP connection for a pair of request/response, which is a very inefficient method.
- To overcome multiple separate connections, two more schemes were introduced – pipelining and multiplexing in the later versions of HTTP protocols. The **Connection** header controls the attributes of the connection partially to make the warm connections.
- The latest version of the HTTP can handle or control the following functions:
  - **Relaxing the origin constraints** – Earlier, the pages from the same origin could access the full information on a web page. This was implemented due to hackers' invasions regarding privacy by using the snooping mechanisms. Now, HTTP separates documents as patchwork from different sources.
  - **Caching** – Caching of the documents can be controlled by the HTTP protocol at different proxies and clients by instructing different conditions of caching in terms of time and other parameters.
  - **Authentication** – By using different schemes such as **WWW-Authenticate** header or cookies, the authentication of users on certain resources can be implemented and controlled by HTTP protocols.
  - **Session control** – The session control can be implemented for certain processes by setting cookies. Cookies can establish a session for a certain process.
  - **Proxy and Tunneling** – Different proxies can be used for accessing the secure servers, which are normally located within a private network without showing the IPs. In such cases, the http protocol uses messages passing through the proxies to reach the main server located on a private network.
- The communication flow of an HTTP protocol is mentioned in the following step-by-step activities performed by the entities of HTTP protocol:
  - Client requests to open a TCP connection
  - Send an HTTP request for fetching a resource
  - Get the response from the server
  - Close the connection OR request to reuse for other requests
- HTTP protocol uses different types of application programming interfaces (APIs). A few of them are listed here:

- XMLHttpRequest API
- Fetch API
- Server-Sent Events
- EventSource
- The XHR API is the core entity for fetching data under the AJAX (Asynchronous JavaScript and XML) and JSON (JavaScript Object Notation) concepts. XML means ""

The additional features and capabilities will continue to grow in the future because the HTTP protocol is extensible and evolving.

Simple Network Management Protocol (SNMP)

Simple Network Management Protocol, precisely known as SNMP, is a network management protocol. This protocol is used for managing a large network of connected devices remotely. This protocol establishes communication between clients, which are pieces of software embedded into the supported network elements such as computers, routers, switches, modems, and other nodes of the network. It is an application-level protocol that runs on a standard TCP/IP protocol suite. It uses the UDP transport layer, which is connectionless and unreliable protocol.

Initially, this protocol was defined by the Internet Architecture Board through RFC 1157 document. Numerous other RFCs have been introduced for the latest versions and upgrades of the SNMP protocol till today. The main purpose of this protocol is to manage a large number of network elements through a centralized manager software developed by different organizations. The main components of a basic structure of SNMP protocol are listed here:[138]

- SNMP manager
- SNMP agents
- Management information base (MIB)
- Managed devices

Three main versions of the SNMP protocol with additional upgrades have been released till today. The latest version of SNMP is known as version 3. All these versions and upgrades were introduced to enhance the features, functions, capabilities, and security measures. The list of the major versions released at the time of writing this book is here:

- **SNMP version 1** – It is a community-based security version.
- **SNMP version 2 c** – It is also an advanced version based on community security.
- **SNMP version 2 u** – This version was introduced for user-based security.
- **SNMP version 2** – This version is focused on the party-based security.
- **SNMP version 3** – It focuses on the user-based security.

SNMP Manager

An SNMP manager is a combination of a computer and a software, which is referred to as Network Management System (NMS). The role of the manager in the system is to establish communication between multiple nodes or network elements in a connected network for monitoring their performance, status, and configuring different parameters on the network elements. The main functions of an SNMP manager are listed here:

- Sending queries to SNMP agents
- Collecting responses from SNMP agents
- Configuring the supported variables in the agents
- Acknowledging the synchronous events from the SNMP agents
- Monitoring of managed devices remotely
- Visual, graphical, and numerical presentation of network, data, and other information
- Correlating different metrics and information within the entire network elements

There are a large number of organizations who deal with the hardware manufacturing of the network elements or nodes such as routers, computers, switches, firewalls, hubs, modems, and other telecom equipment. Hence, there are also several SNMP managers. A few of those popular SNMP managers are listed here:[139]

- **Open View** – Developed by Hewlett-Packard
- **CA-Unicenter** – This NMS is developed by Computer and Associates Inc. commonly known as CA technologies
- **Sun Solstice** – Developed by Sun Microsystems
- **NetView/6000** – Developed by IBM Corporation
- **TME 10 NetView** – Developed by Tivoli Systems
- **And many others** – Developed by many other organizations

The other major features, functions, and characteristics of an SNMP manager are mentioned in the following list:

- It uses UDP so that extra load on the network is not put to create congestion.
- It offers both the graphical user interface (GUI) and command-line CLI interfaces.
- It is also referred to as NMS.
- It can talk to all devices that are SNMP enabled.
- An SNMP manager uses the following different types of messages to request information from the SNMP agents:
  - **GetRequest** – This message originates from the SNMP manager for an SNMP agent connected through the integrated network to send required data or retrieve data from the agent.

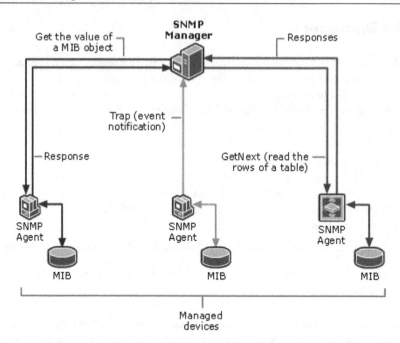

FIGURE 5.6   SNMP manager and agent messages. Here, MIB is Management Information Base (Flickr).

- **GetNextRequest** – This is almost the same message like GetRequest. But this message is used for discovering the available data with the SNMP agent and instructing the agent to keep sending the data with respect to certain terms and conditions.
- **GetBulkRequest** – This message is used by the SNMP manager to retrieve bulk data from the SNMP agent at once.
- **SetRequest** – This request is sent by the SNMP manager to the agent to set certain values and parameters of an object instance for the network operations.
- The SNMP manager's messages sent out to agents and their response originating from the clients are shown in Figure 5.6. The details of the messages used by the agents will be discussed under the sub-topic of "SNMP Agent" shortly.

In the response to the aforementioned messages sent by an SNMP manager to the agent, the SNMP agent generates the responses, which are sent through certain messages. The following sub-topic discusses this issue.

SNMP Agent

An SNMP agent in an SNMP protocol is a software program, which is installed on the network device as an embedded software for connecting the device to the network management system for remote management. In the modern manufacturing of IT devices, it is almost a standard to embed the SNMP agent in the devices at the time of manufacturing with proper ports and other arrangements. The examples of devices that come with the embedded SNMP agent software include computers, mobiles, laptops, routers, switches, modems, signal repeaters, signal amplifiers, microwave systems, transmission equipment, firewalls, hubs, and so on. There are two components of an SNMP agent, which are:

- **Master agent** – also referred to as management definition
- **Subagent** – also referred to as management database

A master agent is a part of an SNMP agent that performs the following activities and responses on the SNMP-powered network management system:

- Provides an interface between subagent software and the SNMP manager
- Parsing of the messages received from the SNMP manager
- Routing of the messages received from the SNMP manager
- Gathering and formatting of the responses retrieved from the subagents
- Sending the response to the SNMP manager
- Notifying the SNMP manager in case when an invalid request is received or data or information required is not available

At the same time, the subagent has also certain dedicated responsibilities in this integrated system of network management. The subagent, which is a part of software of the SNMP agent installed on the devices, handles the following activities and functions:

- Providing the information to master agent against the request from the SNMP manager
- Receiving the requests from the master agent
- Collecting the requested information
- Sending the requested information to the master agent
- Notifying the master agent in case it receives an invalid request or the requested information is not available to provide

To communicate with the SNMP manager, the SNMP agent in the SNMP protocol uses different messages. Other than sending certain messages, the agent has to receive, understand, and act in response to the messages received from the SNMP manager. An agent software uses the following messages to establish meaningful communication within the SNMP protocol.

- **Trap message** – This is a type of message is used by the agent in the SNMP protocol system to inform the network manager about the occurrence of any kind of fault or other uneven events. This message is generated by the agent without any request from the SNMP manager. Normally, agents are configured to send trap messages continuously until the **InformRequest** message is not received from the SNMP manager.

- **Response message** – An SNMP agent has to generate a response message against every request received from the SNMP manager. The content and information in the response message may vary, but the basic format of the message remains the same.
- **InformRequest** – This is the latest type of message, which has been introduced in the latest version of SNMP protocol (v3). This message is generated by the SNMP master in the response to the trap message sent out by the agent. It is just the acknowledgment of the fact that the SNMP manager has received the trap message.

Management Information Base (MIB)

Management Information Base, precisely referred to as MIB, is a database based on certain parameters related to the managed device on which the SNMP agent resides. The MIB consists of a range of fields, which registers a particular information about the status, performance, conditions, and many other factors prevailing on the managed devices that are connected to the integrated network management system, NMS. The MIB is a database, which is shared between the local SNMP agent residing on a managed device and the SNMP server. The MIB database consists of a set of control values and statistical values, which are defined on the managed devices, which embed an SNMP agent interface and software on it. In other words, the MIB database consists of two types of registers or fields as mentioned in the following.

- **Standard fields** – These fields record standard values or parameters of a managed device.
- **Private fields** – These fields are designated for private definition of parameters in any managed device. The SNMP protocol supports multiple fields for customized parameters.

From the SNMP manager perspective, an MIB is a set of questions to query from the agent to get the values of those queried parameters. And from the SNMP agent perspective, MIB is a set of information based on certain parameters (statistical and control values), which are collected from the local managed device and store the data in the field to provide to the SNMP manager on request. A manager should be configured with the full details of the private sets of MIB or questions defined on the managed devices through the SNMP agent residing on the managed device. This enables the SNMP manager to query about the value of that particular parameter defined on the local managed device and manipulate the values to show the presentation on the GUI or generate certain notifications to alert the operations and maintenance teams to look into the matter.

Managed Components or Devices

The managed devices or managed components are the connected devices that are enabled with the SNMP interface and embedded software to be accessed and managed remotely through the NMS powered by the SNMP protocol. The managed devices can be located anywhere in an integrated system connected through a private or public network such as the Internet.

In the earlier days of communication, the managed devices were limited to the major nodes used in the communication. But with the advent of the modern concepts of industrial automation and the IoT, the scope of the managed devices has transformed by 360 degrees. Nowadays, almost all electrical, electronic, and mechanical devices with the capability to be connected and managed remotely are parts of the modern domain of managed devices or the managed components of an integrated remote system.

A few major examples of modern types of managed devices connected through the Internet or other network connectivity with the NMS are listed here:

- Computers, laptops, tablets, and others
- Routers, switches, hubs, firewalls, and others
- Modems, multiplexers, amplifiers, rectifiers, controllers, and others
- Base stations, BSCs, MSCs, mainframes, microcomputers, microwave links, and others
- Printers, copiers, faxes, video phones, security surveillance systems, and others
- Industrial machines, robots, conveyer belts, and others
- And, many types of other home, office, and industrial equipment powered by remote management capabilities through SNMP protocol

It is very important to note that the SNMP protocol is evolving to strengthen the cybersecurity and other features of the protocol so that the systems powered by this protocol can be saved from cyber-attacks on the Internet. The SNMP protocol uses the UDP protocol at the transport layer. The UDP protocol is lightweight; so, it does not put extra load on the bandwidth of the data links. At the same time, the UDP is comparatively easy to exploit for launching cyberattacks. Therefore, more measures are required to make the security more robust and reliable.

The latest version of SNMP protocol (version 3) has additional security features defined in three different security profiles named security levels. Those three security levels introduced in the SNMP protocol v3 are mentioned here:

- **NoAuthNoPriv** – The name of this security level has been derived from (No Authentication, No Privacy). This is the basic level of security, in which the community string is used for the authentication and no encryption method is applied for privacy.
- **AuthNoPriv** – The full description of the name of this security level is (Authentication, No Privacy). In this security level, two methods for authentication –Message-Digest 5 (MD5) and Hash-based Message Authentication Code (HMAC) – are used for authentication. And, no mechanism is used for the maintenance of privacy.
- **AuthPriv** – This name of the security level profile is (Authentication, Privacy). In this profile of security level, the HMAC is used with either MD5 or Secure Hash Algorithm (SHA) for authentication purposes. For privacy maintenance, the DES-56 encryption algorithm is used.

Simple Mail Transfer Protocol (SMTP)

Simple Mail Transfer Protocol, precisely referred to as SMTP, is one of the email handling protocols. It is used for transferring emails between two servers and a pair of servers and clients on the Internet or private network. The SMTP protocol is an application layer protocol, which is fully dependent on the other protocols dealing with the data transmission at the transport and network layers of the OSI model of communication.

The major underlying protocol suite that the SMTP depends upon heavily is the TCP/IP protocol suite, which is also the foundation of the data transmission and routing to the desired destinations as well as networks in the modern Internet environment. The SMTP protocol is also capable of sending the simple documents attached to the emails to the desired email server. The SMTP protocol is a very popular protocol, which is extensively used on a wide range of email servers across the globe.[140]

The basic architecture of an SMTP protocol communication is based on the client and server model. In this model, both the server and client machines should have the SMTP installed for sending electronic mails. The basic schematic architecture of SMTP communication between two email users and the server is shown in Figure 5.7.

In Figure 5.7, the sender uses the UA on the client software for creating and preparing for sending the email message in a proper format. The UA will also send a request to the local Message Transfer Agent (MTA) residing on the local client to establish a TCP connection with the remote MTA residing on the server side of the network. Both MTAs establish a TCP end-to-end connection between the two MTAs and email starts transmitting over the TCP port established between the MTA parts residing on the client and the server. Thus, the email reaches the server at the destination network. This server adds the

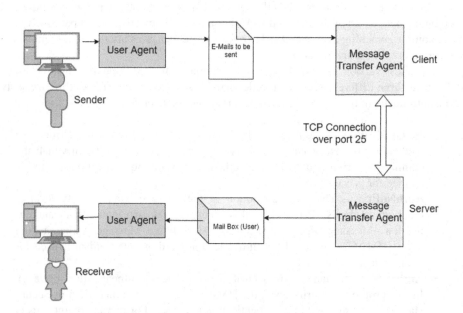

FIGURE 5.7   Schematic architecture of SMTP.

email to the mailbox of the user from where the pulling agent on the receiver terminal will collect the email received from the sender.

The main features, functions, and characteristics of SMTP are mentioned here:

- SMTP protocol is an application layer protocol supporting end-to-end connection over TCP communication protocol.
- This is protocol is based on the client–server model of communication.
- It is an asymmetrical protocol, which communicates with a wide range of clients simultaneously.
- SMTP can both send and receive the emails but is not used extensively for receiving the emails due to issues with the email queuing. For the purpose of receiving the emails, a few other protocols such as Post Office Protocol (POP) and Internet Message Access Protocol (IMAP) are simultaneously used by the mail users or clients. These protocols will be discussed in the coming topics.
- SMTP is extensively referred to as push protocol for emails, and POP and IMAP are known as pull protocols for emails.
- A mailing client wanting to send an email to the server establishes an end-to-end connection through TCP protocol over port number 25 with the help of Mail Transfer Agent (MTA).
- SMTP protocol supports two models of communication as follows:
  - End-to-end model
  - Store and forward model
- In the first model, the client SMTP on the sender side establishes an end-to-end connection with the client SMTP at the receiver end with the help of server SMTP. The server SMTP retains the copy of the email until the message has been transferred to the destination client. It will remove the copy after receiving it at the client end has been completed. The second model is commonly used for the clients within an organization. In this model, the client sends the email to the server, which stores the email, and then sends a copy of the email to the destination client later when the client is ready to receive the email.
- The email server in the SMTP protocol is always in the listening mode. When it receives a request from the client for a TCP connection, it starts a TCP connection over port number 25.
- The SMTP protocol uses two logical entities for mail transfer. They are known as "**User Agent (UA)**" and "**Mail Transfer Agent (MTA)**".
- Both logical components or agents – **MAT** and **UA** – should be available on the client as well as on the server.
- The UA on the sending client prepares and sends the email to the local client MTA, which establishes a connection with the MTA located on the server to send the emails. The UA located on the server checks the mailbox after a certain interval to download the mails in the mailbox to the client.
- To add advanced capabilities in attaching different types of content and other additional information in the email, a few modifications were made in the SMTP in 1993. This advanced version was named Extended Simple Mail Transfer Protocol (ESMTP).

- A few quick revisions were released for the ESMTP in a couple of years after the initial launch and notified by different RFC documents.
- On each mail server powered by SMTP, there is a software agent known as Mail Delivery Agent (MDA). The main purpose of this agent is to deliver the emails received from senders on the server to the designated mailboxes if they are located on that server. If the user mailboxes are not located on that server, then MDA does nothing, but SMTP starts sending the email to the other server, which has the desired user mailbox.
- In most of the cases, the MDAs are built-in within the SMTP's MTA software on the server.
- The SMTP mail transaction consists of two parts as listed in the following:
  - **Envelop** – consisting of header which has sender and receiver addresses and other information
  - **Message** – consisting of the data command and text after the header, which consists of receiver and sender addresses
- The envelop is transmitted separately from the message in SMTP email protocol.
- The envelop command consists of MAIL FROM, RCPT TO, and a null line with <CRLF> to end the command for an envelop sending.
- The mail header consists of the following information:
  - Number of keywords
  - Values defining sending data and time
  - Sender's address
  - Location of the replies
  - And other information
- The communication between the client and the server in the SMTP environment takes place through transactions or dialogues.
- The flow of transactions for delivering an email is mentioned here:
  - Client requests for a TCP connection for the local SMTP, which in turn requests the destination server SMTP to establish a TCP connection. If the server is ready for receiving the SMTP message, it will return 220 code, which means service is ready. If the service is not available, it will return 421 code, which means service is not available.
  - **Server**: 220 domain of server SMTP Service Ready
  - **Client**: HELLO (Client's domain name)
  - **Server**: 250 (Server's domain name). This means server okays receiving the email
  - **Client**: Mail From (sender's Email@domian.com)
  - **Server**: 250 OK
  - **Client**: RCPT TO (Ereceiver's Email@doamin.com)
  - **Server**: 250 OK
  - **Client**: DATA
  - **Server**: 354 Start Mail Input, end with <CRLF>
  - **Client**: Data 30 March 2021 00:34:45 UTC
  - **Client**: From (sender's Email@domian.com)
  - **Client:** Subject Purchase Order

- **Client:** To: Recipient address 1
- **Client:** To: Recipient address 2
- **Client:** CC: Recipient address
- **Client:** (This will be a blank line to inform the server that the body of the message starts from this point)
- **Client:** Copies the entire body of the message
- **Server:** 250 OK
- **Client:** QUIT
- **Server:** 221 (Server's domain name) Closing Transmission Channel
- SMTP protocol uses numerous commands for establishing, managing, and terminating the communication for transferring the emails from the client to the server and finally to the receiving client. A few of those most commonly used commands are mentioned in the following with proper descriptions for developing a better understanding:
  - **HELLO** – This command is used by the client in the SMTP environment followed by its own domain name to get introduced with the mail server powered by the SMTP protocol.
  - **EHLO** – This command is an extended version of the HELLO command.
  - **MAIL FROM** – This command followed by the email address of the sender is sent out by the client to the server for receiving an email from the sender.
  - **RCPT** – This command is augmented by the receiver's email address and is sent by the client to the server to collect mail for that receiver. This can be a single command or can be multiple commands in sequence to introduce multiple recipients to the server.
  - **DATA** – This command is issued by the client to the server for receiving the data in sequential order.
  - **NULL LINE** – A blank line sent by the client informs the server that the header of the email has ended and the message text has started.
  - **Period Line** (.) – A line with just a period sign indicates the server that the body of the message has completed.
  - **QUIT** – This message is issued by the client to terminate the communication channel.
  - **RSET** – This command interrupts the current transactions of the emails and resets the transactions for the email.
  - There are many other commands that are used for requesting more information, defining the delivery conditions, authorization, security, and many other attributes of an email and communication between the server and the client. Most of them are not used very frequently in normal communication. A few of those commands include VRFY, TURN, STARTTLS, SOML, SAML, SEND, NOOP, EXPN, HELP, AUTH, and others.
- Every command is replied with a response, which consists of two parts. One part is the response code and the other one is a short description of the response code. A few important and the most common codes used in this communication are listed here:
  - **220 Domain service ready OR ready to start TLS** – This response is normally generated against the service request by the client.

- **211 System help reply OR system status** – This message is used for getting help about the information sent earlier or informing about the system status.
- **221 Domain service closing transmission channel** – This response code is normally generated against the QUIT command to close the channel.
- **250 OK** – This response is sent by the server (**queuing for node started**) as an acknowledgment of the command issued by the client during mail transferring.
- **354 Start mail input; end with** <CRLF>. <CRLF> – This code informs the client to send mail data and end with the CRLF null line.
- **421 Domain service not available** – It is generated in the response for TCP connection or HELLO message to indicate that service is not available; therefore, closing the transmission channel.
- This is important to note that all response codes are classified into four major categories as mentioned here:
  - Positive completion replies codes
  - Positive immediate replies codes
  - Transient negative completion replies codes
  - Permanent negative completion replies codes

This is very important to note that the SMTP was designed for transferring simple text email messages, which abide by the rules of ASCII code. Therefore, the power of SMTP was limited, and additional capabilities were required to support other coding systems for creating messages and also sending the data in other formats such as video, audio, binary, images, and any other format. To fill this capability gap, a new protocol was developed which is used as an extension of the SMTP protocol to add the required capabilities. That protocol is known as Multipurpose Internet Mail Extension (MIME) protocol. This protocol is described with full details as the following topic.

Multipurpose Internet Mail Extensions (MIME)

Multipurpose Internet Mail Extensions, precisely MIME, is a type of protocol that defines different forms of texts, coding systems, formats of data, and many other features of data, which are required to be transferred through emails over the SMTP and other email protocols. As the name indicates, it is an extension, which works over the base email protocol.

Primarily, this protocol was designed for the SMTP protocol, but other email as well as communication protocols also benefited from the definitions of the data formats, encoding, and other formats. The data formats by the MIME protocol are extensively used in web-based communication protocols such as HTTP and HTTPS. Nowadays, it is extensively supported by all major email protocols.

SMTP protocol was designed as a plain text protocol, which supports only ASCII text coding for messages. Any language or text format other than ASCII code was not possible to be sent over the SMTP protocol. Thus, there were a large number of limitations of SMTP protocol to support other languages commonly used in the world such as French, German, Spanish, and others. Meanwhile, different formats of attachments such as documents, images, audio, video, executable binary files, and other data formatted in different coding schemes were not supported by the SMTP.

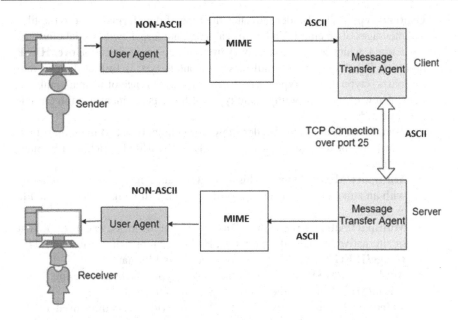

FIGURE 5.8 Message code conversion and transmission flow by MIME.

To address this issue, the MIME extension was designed by an engineer Nathaniel S. Borenstein who was working at Bell Communications (Then Bellcore Inc.) in 1991.[141] The main purpose of developing this extension protocol was to offer benefits of SMTP protocol to the people that speak different types of languages, which are not digitally coded in ASCII standard. The message conversion and transfer flow powered by the MIME extension is shown in Figure 5.8.

The working principle of MIME is very simple. When a UA wants to send an email message that is not in ASCII code format, the MIME converts it into the 7-bit NVT ASCII (Network Virtual Terminal ASCII) code and adds the information header to the message and sends it to the destination. It adds different headers to the entire email message before sending it to the email server. After converting the message from a non-compatible code to the compatible one (ASCII code), and adding the headers to the messages, a message is transmitted over the network with the help of SMTP protocol. At the receiving end, the MIME protocol recovers the message into the original format to present to the end-user through a UA software tool.

There are five major types of headers, which are added to the email message. Those main headers used in the MIME protocol are mentioned here:

- **MIME Version** – This header shows what version of MIME has been used in the message encoding. Due to some technical limitations, the second version of the MIME was not introduced. But, an additional release of MIME was released after version 1.0 in the initial launch named MIME version 1.1. The second release of the MIME extension covers the limitations related to supported word count and other capabilities of the SMTP protocol and related communication protocols used in this communication.

- **Content-Type** – This header indicates the type of data format used in the files and messages of an email. Under this category, numerous types of data formats are defined, which are also extensively used in web communication over HTTP protocol. The definition of email message content type is further divided into two parts – type and subtype parts – to cover a wide range of format combinations. A few of those very important types of formats are mentioned in the following list:[142]
    - **Text/Plain Type** – This header shows the content type text in plain format. Similarly, the text can be in Text/HTML and Text/RTF (Rich Text Format) formats.
    - **Multipart/Mixed Type** – This indicates that content type is text along with an attachment in a certain format, which will be indicated by the file extension.
    - **Multipart/Alternative Type** – This shows that the message content is sent in alternative formats like plain text, HTML text, or rich text format RTF.
    - **Image/JPEG Type** – Shows the type of file and format.
    - **Application/MSWord** – Shows the content type attached in the file.
    - **Audio/MP3 Type** – Indicates the type of file attached.
    - **Video/MP4 Type** – Describes the type of video encoding used in the content.
- **Content Type Encoding** – This header defines the type of coding used for the binary-to-text code in the encoding of the text and other data in the message of an email. The main types of encodings used include:
    - 8-bit encoding
    - 7-bit encoding.
- **Content Description** – This header specifies the class of content in the body of the message whether the content is video, text, or other kinds of content through a short description in this MIME header.
- **Content ID** – This header specifies the unique ID of the message.
    After having discussed the main headers and functionality flow of the MIME extension working in parallel with the SMTP protocol, let us figure out and summarize the main characteristics and features of MIME extension with the following list:
- MIME extension works as a complementary protocol above the SMTP email protocol as an extension to it.
- Primarily, the Pretty Good Privacy digital signature and privacy was used for the security purpose. This scheme used public key cryptography for security.
- Later on, IETF standardized new security capabilities to the MIME referred to as S/MIME, which was developed by the RSA Security. This security mechanism adds secured sections into the MIME extension for encrypted emails. This section is added as "Security Multi-parts for MIME". These multi-parts for MIME include the following types:
    - Multipart/Encrypted
    - Multipart/Signed.
- MIME enables the SMTP protocol to send a wide range of formats of data with proper security, privacy, and integrity.

- MIME overcomes the limitation of message word count and allows to send an unlimited number of characters through a mail message.
- Allows sending executable files over the SMTP. Those files were not supported in the simple SMTP protocol.
- MIME can be used along with POP, SMTP, and other email access protocols.
- The types of contents defined by the MIME are used by the web communication protocol, HTTP over the www-based Internet.
- MIME enables the SMTP to transfer multiple attachments in an email message.
- Supports different fonts, layouts, and colors in an email message.
- MIME also allows the developers and network administrators to register for a new content type for customized transmission of data.
- MIME extension supports other additional functions as listed here:
  - **Signed data** – encrypted through base-64 encryption
  - **Enveloped data** – through the receiver public key
  - **Singed and enveloped data** – the combination of the aforementioned functions
  - **Security label** – supports the level of security through different labels
  - **Secure mailing list** – A trusted list of recipients.
- MIME extension uses the following types of encryption algorithms:
  - **RSA** – for digital signature and symmetric key encryption
  - **Diffie Hellman algorithm** – for symmetric key encryption
  - **DES-3** – for encryption with symmetric key
  - **DSS** – for the purpose of digital signature.
- MIME extension protocol uses X.509V3 digital certificate.

File Transfer Protocol (FTP)

File Transfer Protocol or precisely referred to as FTP, is a protocol for transferring the files of data from one computer to another computer. This protocol is an application layer protocol like HTTP, Telnet, and others. The communication through FTP protocol is based on the server and client approach. A client requests the FTP server to provide access and then transfer the requested files to the client. The communication between the server and the client occurs on the two parallel connections. One connection is used for controlling the communication and is referred to as *control connection* and the other one is used for transferring the data and is known as *data connection*. Before we deep-dive into the technical aspects of the protocol, let us have a look at different aspects of FTP that are concerned with the common users in the networked environment, especially in the Internet ecosystem that every connected user uses on a regular basis.

Any file located on a server computer can be accessed in multiple ways using the FTP protocol as listed here:

- Using web browser interface
- Using the command line interface
- Using FTP client software tool

Every way of accessing and downloading the data from an FTP server uses a separate method for accessing and downloading the data on the local computer. Accessing files through FTP client software tools is a very popular and efficient way and is nowadays used by a large number of normal users as well as by technical users. A schematic diagram of the entire communication flow of the FTP function is shown in Figure 5.9.

Both data connection and control connections use the TCP/IP protocol suite for transporting the data and control over the Internet. The control connection handles the control process at the client as well as at the server end. On the other hand, the data connection accomplishes the data transfer processes at both the server and client ends. This process also directly deals with the local data disks or storage at both server and client machines. This is very important to note that FTP servers always use the out-of-band control connection, but in some web-based application layer communication protocols like HTTP and SMTP, the same in-band connections are used for transferring both the control commands and data over the same communication channel. They are known as in-band communication connections and FTP connections are also known as out-of-band control (connections).

The main features and characteristics of FTP are mentioned in the following list:[143]

- The FTP connection is initiated by the client, which may be a browser, an FTP client software, or a command through a command prompt.
- A close competitor of FTP protocol is HTTP protocol, which is also used for transferring the data files on the Internet. But the HTTP has some limitations in terms of speed, authorization, and other factors.
- A client can request to establish a data connection in two major modes:[144]

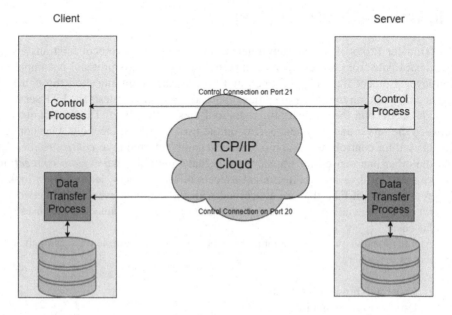

FIGURE 5.9   FTP communication schematic diagram.

- **Active connection mode** – In this mode of connection, the client requests the server to establish a control connection on a random port designated by the client – for instance, port number 1. The request is sent through the "**PORT**" command issued by the client. The server responds to the request by opening a connection on port number 21 and the designated port number by the client. Then, the client requests for a data connection on another random designated port number – for example, port number 2. The server responds by opening a data connection on port number 20 with the designated port number 2 by the client. This mode is normally used when there is no firewall enabled on the client side; otherwise, the passive mode is used.
- **Passive connection mode** – In the passive mode, the user designates the port initially and sends a request through the FTP client for establishing an FTP connection on the random port. The FTP client requests an FTP connection with the server on port 21, but the server sends a command named "PASV", which carries the random port designated by the server to open for an FTP connection on the client – for example, port 3 or any other port. The connection is established for data connection through that designated port by the FTP server and data transfer starts.
- The data connection is established for the transfer of one data file. After the completion of the transfer of that data, the data connection is terminated, but the control connection is not terminated and a new request for the establishment of another data connection can be negotiated through the control connection.
- Multiple files and folders can be transferred by using the FTP protocol between two machines.
- The data transfer through FTP is much faster than other communication modes such as HTTP or even other legacy forms of data transfer.
- FTP protocol is a *stateful* protocol, which maintains the state of the user connection during the entire communication session.
- Supports very large files and folders to be transferred.
- FTP protocol does not encrypt the data during transfer, which may pose some security risks to the data during transmission.
- Numerous FTP clients are commercially available in the market such as FileZilla, FireFTP, CyberDuck, WinSCP, and others.
- FTP uses the TCP/IP suite of protocols at the lower layers of the OSI communication model.
- FTP server that is for public use and does not impose authentication and authorization is known as an "**Anonymous**" server. Otherwise, a client will need to provide username and password for accessing the data directories on the server.
- FTP protocol supports both structured and unstructured data files.
- The FTP protocol supports three types of data structures of the stored files on the server as mentioned here:
  - **Page structure** – In this type of data structure of file, the file consists of independent indexed pages on the server.
  - **File structure** – In file structure type, the data file is a continuous sequence of data bytes without any internal structure.

- **Record structure** – In this kind of data structure, the file is constituted by the sequential records of data.
- FTP protocol uses a wide range of commands for communication between server and client. A few most common commands used in the FTP protocol for establishing, managing, and terminating the control and data connections are mentioned here:
    - **PORT** – Command is used to instruct the server to designate the port for the control connection.
    - **PASV** – This command is used to establish a passive mode of data connection.
    - **USER** – For sending username information for authorization.
    - **PASS** – For sending password information for authentication.
    - **RETR** – This command is used for establishing a data connection with the remote host and transferring the data over that data connection.
    - **STOR** – It is used to store the data file on a remote host in the existing directory.
    - **LIST** – This command is used to display all files available in a directory.
    - **QUIT** – This command is issued to terminate the user connection if the file is not in process of transferring.
- As we know, the FTP protocol is based on the client/server model of communication. The command is issued in this communication, which is acknowledged or responded with a response, which is known as replies. Those responses are provided in the combination of response code and a short description. A few very common and most important responses are mentioned here:
    - **200 Command OK** – This response is issued in the response of every correct command received and implemented by the remote host.
    - **225 Data Connection Open; No transfer in progress** – This response is sent when a new connection is requested despite the control connection exists.
    - **221 service closing control connection** – Normally responded against the QUIT command to close the user connection.
    - **331 user-name ok; need password** – This response is issued against the correct USER command to seek the password.
    - **502 command not implemented** – Generated by the remote host in the cases when the issued command is not implemented due to certain reasons.
    - And many others.
- FTP protocol supports three types of data presentations as listed here:
    - ASCII 7-bit coding
    - EBCDIC 8-bit coding
    - 8-bit binary data coding.
- FTP communication supports the following-mentioned three modes of transmission:
    - **Stream mode** – This is the default mode of transmission in FTP protocol. In this transmission mode, the data is transferred in the shape of stream without any segmentation of the data. The segmentation is handled by the TCP protocol.

- **Block mode** – In this mode, the data is distributed in blocks and sent to TCP in blocks. A 3-byte header, referred to as block header, is added. One byte is block descriptor and two bytes specify the size of the block.
- **Compressed mode** – In this mode of transmission, the data is compressed to reduce the size by using Run-Length coding (Mostly). In this coding, the null characters in the binary file are compressed and spaces are removed in the text files before transmitting.
- The control connection is managed by the control process, which is also known as Protocol Interpreter (PI) in the FTP communication.
- PIs are located on both the client and the server. They communicate by using the (network virtual terminal) NVT syntax. Any command in the form of UNIX and DOS is translated into NVT syntax by the PI.

Trivial File Transfer Protocol (TFTP)

Trivial File Transfer Protocol is precisely referred to as TFTP protocol. It is a miniaturized version of the FTP protocol. It has reduced features and capabilities as compared to the FTP protocol. But, the main purpose of this protocol was also the same as that of the FTP protocol. The only difference between FTP and TFTP is the lightweight and capabilities. Before deep-diving into the software-level technicalities of this protocol, let us explain a few most common characteristics of this protocol as mentioned in the following list:[145]

- TFTP protocol is used to transfer the data files between the two hosts.
- It does not support any kinds of modification and changes in the directory of the files as the FTP protocol can do.
- It is termed as the simplest form of FTP protocol for file transfer.
- TFTP protocol is not designed to support authorization and authentication.
- The transporting of the data goes in plain/clear text.
- Cisco Systems uses this protocol on all networking devices, especially the routers and firewalls for backing up and restoration of the IOS images.
- The size of the TFTP software is much smaller than FTP.
- It is extensively used in embedded system software, especially used in the domain of the IoT field.
- Being a small software in size, it is installed on the read-only memory (ROM) of embedded systems to copy the image of the operating system for those devices when the power is turned on of those devices.
- TFTP protocol uses the UDP port number 69, on which it runs.
- It uses retransmission and timeout functions to make sure that data is received by the desired remote host because it uses UDP, which is not a reliable protocol.
- TFTP uses the fixed size of blocks of data for transmission. The fixed size is 512 bytes.
- TFTP ensures the receipt of the first block of 512 bytes before sending the second one.
- The receiver of the block of 512 bytes sends acknowledgment after receiving every block.

- The first packet is of greater importance in terms of establishing communication between two remote hosts. The first packet specifies the following things:
  - Requests for a file transfer to the remote host
  - Establishes interactive communication
  - Specifies the file name
  - Specifies if the file will be written, transferred to the server or read, transferred to the client.
- The block numbers are given in a sequential order starting from 1.
- A header of 3 bytes is attached to each data packet. In the header, the number of blocks contained in a data packet is specified.
- Any block of size less than 512 bytes is treated as the end of the file by the remote host.
- Any kind of error message is received by the sending host, the transmission of data over the TFTP protocol is terminated. But retransmission occurs only in case no acknowledgment is received after a timeout period.
- TFTP protocol uses five different types of messages for establishing, maintaining, and terminating the communication between two remote hosts. The format of those five different messages is different except the one common field, which is the operation code (opcode).
- The details of those five messages are mentioned here:[146]
  - **Read Request (RRQ)** – This message is sent by the client to get a copy of the desired file located on the remote host/server. This message consists of one fixed field known as operation code (Opcode), which consists of 2 bytes. One "Filename" field of variable length, one mode field of variable size. The "Mode" specifies the type of data such as binary, ASCII, or mail. Two variable fields – Mode and Filename – are separated by a one-byte field known as "zero" field. The purpose of this field is to separate two fields. One byte field is also a zero field to separate this message from other communication. The details of the RRQ message fields with their respective sizes are shown in Figure 5.10.
  - **Write Request (WRQ)** – This TFTP message is known as operation code 2 and is used to copy a file on the remote host computer. This message has also five fields as shown in Figure 5.11.
  - **DATA** – This message is used to request transfer data of the requested file. This message consists of three fields – operation code, sequence number, and data. The data field can be either a 512-byte field or any other length ranging from 0 to 511 bytes. All messages except the last message of the data file will have a data field of 512 bytes. The last field will be less than 512 bytes, which is an indication to the remote receiver that the file has completed transferring. The details of the fields of this message are shown in Figure 5.12.
  - **Acknowledgment (ACK)** – This message is used for acknowledging the receipt of the data block through a sequence number of the received data and expected new sequence number of the data block. It is a two-field message – operation code type and sequence number – for sending an acknowledgment. The details of the fields of this TFTP message are shown in Figure 5.13.

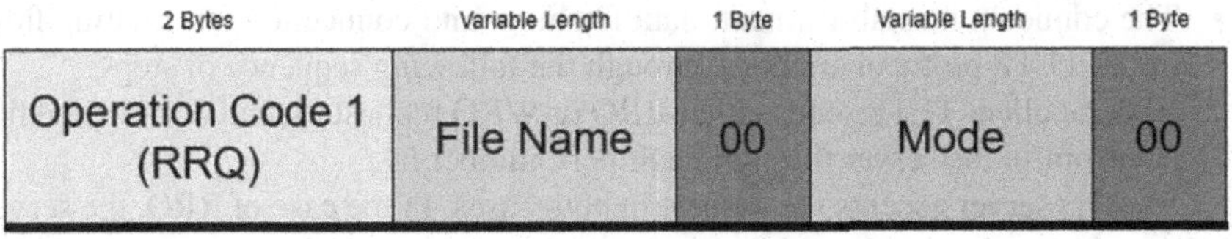

2 Bytes	Variable Length	1 Byte	Variable Length	1 Byte
Operation Code 1 (RRQ)	File Name	00	Mode	00

FIGURE 5.10 RRQ message fields.

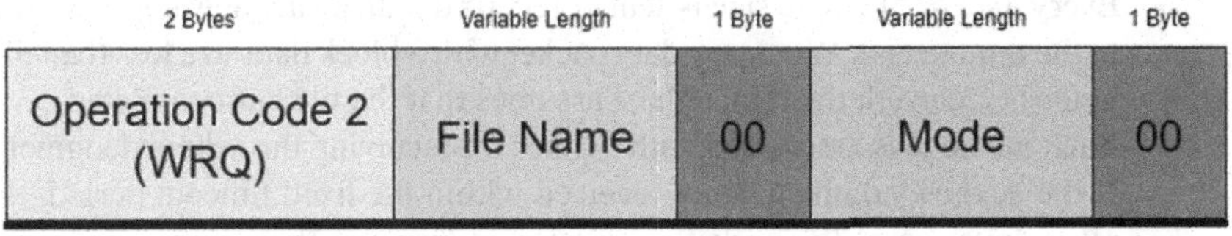

2 Bytes	Variable Length	1 Byte	Variable Length	1 Byte
Operation Code 2 (WRQ)	File Name	00	Mode	00

FIGURE 5.11 WRQ message fields.

2 Bytes	2 Bytes	0 to 512 Bytes
OPCODE 3 DATA	Sequence Number	File data

FIGURE 5.12 DATA message fields.

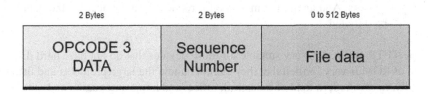

2 Bytes	2 Bytes
OPCODE 4 (ACK)	Sequence Number

FIGURE 5.13 ACK message fields.

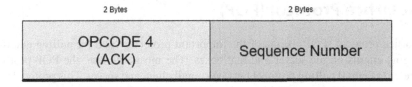

2 Bytes	2 Bytes	N Bytes	1 Byte
OPCODE 5 ERROR	Error code	Error message	0

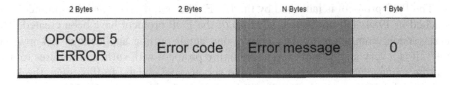

FIGURE 5.14 ERROR message fields.

- **ERROR** – An error message is generated due to failure of the transmission of the file during the fetching operation or an error can occur at the time of sending the first message in the form of RRQ and WRQ requests. When an error message is received, the host stops the operation for any further proceedings. The details of an ERROR message are shown in Figure 5.14.

- The connection establishment, data fetching, and connection terminating flow in the TFTP protocol are done through the following sequence of steps:
  - The client TFTP sends either RRQ or WRQ request to read or write the file from/to the server through UDP port number 69.
  - The server accepts the request in both cases. In the case of RRQ, the server sends a data packet with complete information and sequence number in the data frame. And, in the case of WRQ, the server sends an acknowledgment message to the client over a randomly selected UDP port number.
  - Every data of 512-byte size is sent or received with an acknowledgment sent to the remote end. When any data packet with a block data size less than 512 bytes is received, the remote host assumes that the file has completed.
  - Each message is associated with a timer for receiving the acknowledgment. If the acknowledgment is not received within the fixed timeout period, the retransmission of the packet is initiated in terms of the previous sequence number for the received block of data, which was acknowledged earlier.
  - Any error message would lead to the discontinuation of the file transferring process. And again, the message for requesting a connection and data will be initiated.

The TFTP protocol is very suitable for those devices that do not have hard disks and the size of ROM is very limited that they cannot handle the large protocols and firmware, especially, in the modern field of industrial, office, and home automation and other forms of process automation in the connected world like smart cities and others.

Post Office Protocol (POP)

Post office protocol (POP) is one of the important protocols used for pulling emails and managing emails on the server and mailboxes. The most common, the POP protocol is referred to as email pulling protocol between email client and server. This protocol is configured on the email client software such as MS Outlook, Thunderbird, and others. Those clients use the POP protocol to communicate with the email servers for accessing emails and managing the emails in the mailbox.[147]

The POP protocol is launched by the IETF. The initial version of this protocol was launched in 1984. Till now, three versions of the POP protocol have been created. The latest version, commonly used in present-day email pulling applications is known as POP3. It is a very simple and easy-to-configure protocol with sufficient features to deal with the management of emails within the servers and clients. In this topic, the latest version of the POP protocol (POP3) will be discussed with full details of its commands and responses.

The POP3 protocol works over the TCP/IP protocol suite, which is the backbone protocol suite of the modern Internet and other major types of networks used in Intranet communication. Normally, this protocol uses the TCP port number 110 for connecting to the POP3 server. With advanced capabilities included in the POP3 version, it can provide better security through major types of encryption mechanisms such as Transport Layer Security (TLS) and Secure Socket Layer (SSL) over the secure version of POP protocol commonly referred to as POP3S. The POP3S protocol uses the TCP port number 995 for connecting with the server on a secure and encrypted connection.

Initially, the POP protocol was introduced by IETF standard organization through RFC 918 in 1984. To realize the advantages of reading the emails offline as well as online, more advancements in the POP protocol continued. The revised protocol named Post Office Protocol Version 2, precisely POP2, was introduced in 1985 with Request for Comments, RFC937. This protocol was enhanced very soon and was replaced by the latest version of the protocol known as POP3. This version of the POP protocol was released in 1988 through an RFC1081. For many coming years, this version of the POP protocol was continuously improved and refined till it was finalized in 1996 through RFC1939. [148]

The communication between the POP3 client and server takes place through different commands issued between the server and the client. To check and retrieve the emails from the email server, the client software sends a request to connect for emails. The server asks for authentication through username and passwords. The username and passwords are the same used in the mailboxes on the email server. After verification/authorization, the client connects to the server and starts issuing different commands for retrieving, deleting, listing, and other activities to the server. The server responds with the appropriate action and acknowledgments through certain response codes, which will be discussed later in this topic. In idle mode, the server is always in waiting mode for establishing the TCP connection on the request of the client.

The main features and characteristics of the POP3 email pulling protocol are mentioned in the following list:

- This protocol is based on server–client communication mode.
- The server remains in waiting/listening mode to establish a connection over TCP designated port number on the request of the client.
- A client-host makes use of the POP3 service, and a server-host offers the POP3 services.
- The server listens to the request for POP3 service on TCP port number 110. When a connection is established between client and server, the communication between the two entities starts through different commands.
- Each command is a case-sensitive keyword followed by one or more than one argument and it terminates with a pair of Carriage Return Line Feed (CRLF). The details of the major commands used in the POP3 communication will be discussed shortly.
- Every command is responded with a response specified with a **+OK** or a **-ERR** and the details and description of the parameters requested in the command. The details of the responses generated against the commands issued by the client will also be discussed shortly.
- POP3 communication uses three types of status indicators for generating the response against the commands. Those two status indicators are – Positive (+OK) and Negative (-ERR). The positive responses are generated against the successful completion of the request received through POP3 commands to the server and the negative responses are generated against the commands that have either failed, required resources not available, or any other error occurs during the execution of the command.

- The communication between the server and the client in the POP3 protocol can be classified into four categories of sessions as mentioned here:
  - Port opening and greeting state
  - Authorization state
  - Transaction state
  - Update state.
- A connected server in the POP3 protocol terminates the connection after a certain period without sending any information through response to the client. It is an automated disconnection from the connectivity to save resources.
- The authorization state can be accomplished through two different sets of commands as listed here:
  - Using USER and PASS command combination
  - Using APOP command (to be explained shortly).
- The most common commands used in the POP3 communication are mentioned in the following list with full details:
  - **STAT** – This command is used by the client to know about the status of the mailbox that how many emails are there and what is the size of the mailbox. The server sends a positive +OK response with the details such as the number of emails and the size of the mailbox. The example is given here:

Client: **STAT**

Server: **+OK 4 480**

The server response (+OK) indicates that the command was successfully implemented, and after a space, the number (4) shows that four emails are available in the mailbox and the last number (480) indicates the size of the mailbox in bytes (octets).

  - **LIST** – This command is used to request the server to provide the details of the messages available in the mailbox. This command can be issued with an argument such as email message number or without any argument. The details of the response generated from the server against the LIST command with argument and without argument are mentioned here.

Client: **LIST**　　　　(without argument)

Server: **+OK 3 messages (600 octets)**

Server: **1 150**　　　(The first email is of size of 150 octet)

Server: **2 250**　　　(The second email is of 250-octet size)

Server: **3 200**　　　(The third email is of 200-octet size)

Server: **(.)**　　　　(This line marks the end of the response)

Now let's explain the LIST command with argument:

Client: **LIST 3**　　　(With argument of email number)

Server: **+OK 3 200**　　(Shows the details of email number 3)

  - **RETR** – This message is used to retrieve a particular email message. You need to use one argument, which is email number to retrieve a particular message. If that particular message is not available, the negative response with -ERR will be generated and if that email number is available, the entire email will be retrieved and presented as shown in the following command and response:

Client: **RETR 3**　　(With one argument 3)

Server: **+OK 200 octet**　　　(The email size is 200 octet)

Server: **<Entire message will be printed here>**
- **DELE** – This command is used to delete a message from the mailbox. This command is used with an argument, which is the number of the email message as mentioned here:

Client: **DELE 3**   (Delete message number 3)

Server: **+OK message 3 deleted**
- **RSET** – This command is used to restore the mailbox by undeleting the deleted items. It is used without argument as the command and response shown in the following:

Client: **RSET**   (Reset mailbox without any argument command)

Server: **+OK mailbox has 3 messages (600 octet)**
- **NOOP** – This command is used to instruct the server not to do anything. The server replies with a +OK as shown in the following:

Client: **NOOP**   (Take no action command without any argument)

Server: **+OK**
- **QUIT** – This command is used to delete the messages in the mailbox and disconnect the connection by signing out. The details are shown in the following.

Client: **QUIT**   (Delete messages and sign-out)

Server: **+OK DEWEY POP3 server signing off (maildrop empty)**
- **TOP** – This command is used with message number and number of lines of a message after the header of the email message. This command is designed for getting top lines of a message.
- **UIDL** – This message is used to display the list of unique IDs of email messages.
- **USER** – This command is sent to the server with the name of the user as an argument. The response may be +OK in case the user is real and -ERR in case the user is not available on the server.
- **PASS** – This command with the password as an argument is used for authentication in the POP3 email protocol.
- **APOP** – This command is used for authentication of the user through two arguments – username of the mailbox and the shared MD5 secret. This is used to safeguard the frequent logins through user and password to be compromised in the plain text format over the network.

TELNET Protocol

TELNET protocol is one of the oldest communication protocols. The main purpose of developing this communication protocol was to establish an access to the remote host computer or server. Telnet was introduced in 1969. It supports the byte-based bidirectional communication through plain text transmitted over the network. This makes it prone to hackers' attack to steal the username/password and other information traveling over the connection without any encryption. This was developed about half a century ago; hence, it could be understood that there was no Internet like that we use today. At that time, this protocol was very useful and powerful for accessing the remote hosts or servers for configuration or any other activities to perform remotely.[149]

The word, TELNET is the combination of **Tele**type **Net**work. It is also referred to as the combination of the letters taken from the **Terminal Net**work. This protocol is standardized under the recommendations of ISO under RFC 854. The TELNET protocol works on the basis of the TCP/IP protocol suite. It uses the TCP port number 23 for establishing a connection between two remote hosts. The data as well as the control information travels through the same communication channel established through the designated port number of the TCP protocol. The communication is fully text-oriented. The syntax of the command to access the remote hot is based on three components (Command + Hostname/ IP address + Port number).

The main features and characteristics of the TELNET protocol are mentioned in the following list with full details:[150]

- TELNET protocol is a server/client-based communication protocol to connect a local computer with a remote computer.
- It is an application layer protocol that works over the TCP/IP suite of protocols.
- The telnet client software should be installed/enabled on the local machine, and the telnet server software/service should be installed/enabled on the remote host or server.
- TELNET protocol is managed through the commands issued through a local terminal to the remote server.
- A TELNET command consists of three parts as listed in the following and the format is shown in the schematic diagram shown in Figure 5.15:
  - IAC – Interpret as Command
  - Command code (0 through 255)
  - Option code (a short descriptive code)
- This protocol is built on the basis of the three main considerations or ideas as listed here:
  - Network Virtual Terminal (NVT) concept
  - Symmetrical views of terminals and processes
  - Negotiated options' concept
- The communication between the two terminals client and server takes over the TCP connection in NVT language, which is a format of input character.
- NVT supports two types of character sets known as:
  - Standard format for data characters
  - Standard format for control characters.
- A large number of control functions are performed/accomplished in the TELNET communication. A few very important ones are mentioned here:
  - **Interrupt Process (IP)** – This function is used for terminating, aborting, and suspending the processes or function within the communication.

Interpret as Command (IAC)	Command Code (CC)	Option Code (OC)

FIGURE 5.15   TELNET command format.

- **Abort Output (AO)** – This control function allows the process to complete without sending the output data to the terminal.
- **ARE You There (AYT)** – Used to check whether the system is running or not?
- **DON'T** – This control command is used for the denial of any request in the telnet communication.
- **DO** – This control command is used to agree on a specific option.
- **Go Ahead (GA)** – To continue the process, this command is issued.
- **Start of Sub-negotiation (SB)** – This command is used for the starting of the sub-negotiation option.
- All of the aforementioned commands can only be executed as commands if before that command, the IAC character or function should be sent to the remote host. Otherwise, the command will not be treated as command to execute. The IAC code is 255 as mentioned in earlier in this topic.
- The control commands such as DO, DON'T, WILL, and WONT are used for the negotiation of options in the Telnet communication.
- The major options that are negotiated between the two terminals or hosts in the TELNET communication are listed here:
  - Echo modes
  - Default mode
  - Character sets
  - Line mode
  - Character mode
  - Others
- To negotiate for the other parameters like window size, type of terminal, and other such parameters, the option of sub-negotiation commands is used in the telnetpowered communication between two remote hosts.

Extensible Messaging and Presence Protocol (XMPP)

Extensible Messaging and Presence Protocol, precisely XMPP protocol, is an open-source protocol for instant messaging (IM) communication over near-to-real-time applications. It was initially developed and named as Jabber protocol (introduced in 1999), but later on, it was renamed the XMPP protocol (here, we will use both names interchangeably). This communication protocol for IM is also capable of showing the status of the communication node or partner for providing real-time information about the status of the communication partner. The presentation of the status of the communication partner is known as presence information. The main purposes of this open-source IM protocol include: [151]

- Instant messaging (IM)
- Presence information (PI)
- Contact list maintenance

This protocol is based on the XML. The capability of this protocol is based on the principle of extensibility, which allows the addition of futuristic features and capabilities. It supports numerous middleware programs associated with the IM and other applications. It can support signaling for gaming applications, video, VoIP, and file transfer applications easily.

The number of users of this protocol reached a whopping 10 million plus by 2003 after its initial introduction in 1999. In 2004, the protocol was formally released as a standard for IM in an open-source license scheme. This protocol is almost of the same level and standard as of the SMTP protocol for email communication. A large number of XMPP-powered client software tools are available that support a wide range of operating systems including Windows, Linux, iOS, and others mobile and desktop platforms.

The main features, characteristics, and capabilities of the XMPP IM protocol are mentioned here:[152]

- XMPP is a client–server-based protocol for IM applications.
- It is an IETF-recommended protocol.
- This protocol is an application layer protocol that works over the TCP/IP protocol suite.
- The communication between the server and the client is based on decentralized architecture/structure.
- It is an open-source platform and free to use for a wide range of applications including commercial use and modifications.
- A public domain server run by Jabber protocol is also available for free registration and communication. If somebody wants to run the XMPP server in his/her own organization domain, he/she can do that easily without any restrictions.
- The addressing scheme of XMPP is similar to the email addresses. The addressing ID is known as Jabber ID or JID.
- It also supports HTTP transport for a wide range of web client applications.
- QoS is not supported by this protocol.
- The XMPP model uses the HTTP protocol in two ways – polling and binging. The polling method proved as a bulky solution and wastage of resources; so, it was deprecated then. Now, the binding method is used in HTTP-based XMPP instant message communication.
- The binding method triggers the servers to push the messages when they have been received at the server over a bi-directional stream powered by the synchronous HTTP (BOSH).
- The use of XMPP is continuously evolving due to its extensible nature. It is used in chatting, chat rooms, collaborative tools, file sharing systems, geolocation services, content syndication, network management, cloud computing, and many other environments and applications.
- XMPP protocol is capable of supporting peer-to-peer and machine-to-machine communication across the cross-platform networks.
- Supports the Multi-User-Chat (MUC) conferencing applications over IM and presence information applications.
- This protocol is capable of connecting with other servers using different protocols through gateway servers. This enables the interoperability between multiple protocols used by other IM applications. Any single client application for IM will be able to communicate across the IM and other IM servers without any problems.

- A few examples of client software powered by XMPP include, Swift.IM, Gajim, Converse.js, Monal, and others.
- A few popular server software powered by XMPP include Prosody, eJabberd, and others.
- The entire communication between the two or more than two users of IM and presence information takes place through different types of XML messages among the following entities:
  - C = Client
  - CC = Contact's client
  - CS = Contact's server
  - S = Server
  - US = User's server
  - UC = User's client

The XMPP protocol has become one of the most popular protocols for a wide range of IM applications like the SMTP has become a very popular protocol for email applications in a large number of private, public, and commercial email applications.

Jingle Protocol

Jingle protocol is a complementary extension to the XMPP protocol to initiate, manage, and tear down a media connection between two entities or applications powered by the XMPP IM protocol. The main objective of developing this new extension for establishing a peer-to-peer media channel was to simplify the use of existing protocols for messaging communication like XMPP. Adding new features to the XMPP that can interoperate with other real-time communication protocols is a bit complex and would incur extra burden on the communication channels and systems, which would lead to degradation in the performance of the system.[153]

The main features and characteristics of the Jingle extension or protocol are mentioned in the following list;[154]

- An XMPP extension is to add session control signaling for managing peer-to-peer multimedia channels such as video conferencing, VoIP, and others – between two XMPP-powered communication elements.
- Developed by XMPP Standard Foundation and Google Inc.
- Real-time Transport Protocol (RTP) is used for delivering multimedia streams.
- *libjingle* library has been released for Jingle implementation by Google under BSD license.
- A wide range of clients support the Jingle. A few of them are listed here:
  - FreeSwitch
  - Asterisk PBX
  - Google Talk
  - Gajim
  - iChat.

- Modular architecture of Jingle extension supports a wide range of multimedia apps, transport methods, and security modules with the help of developers.
- Session establishment, management, and tearing down take place over the XMPP protocol and multimedia transmission out of the XMPP communication.

WHAT ARE ROUTING PROTOCOLS?

Routing protocols are the most important protocols used in the routing of the data packets or signaling information from one network to another network. Routing protocols are the sets of rules to transport the data from one router to another across multiple communication networks interconnected within a network of the networks.[155]

There are different types of protocols used in routing the data packets from one router to another. The most important types of routing protocols are associated with IP routing. The IP routing protocols can be classified into numerous categories.

MAJOR CATEGORIES OF ROUTING PROTOCOLS

The routing protocols can be classified in terms of numerous characteristics, their uses, and attributes. A few major categories are listed here:

- **Purpose-Based Protocols** – Such as Exterior Gateway Protocol (EGP) and Interior Gateway Protocol (IGP)
- **Behavior-Based Protocols** – Such as traditional *classful* protocols and modern *classless* protocols
- **Operation-Based Protocols** – Such as path-vector protocols, link-state protocols, and distance vector protocols

The brief definitions of these different categories of routing protocols are mentioned in the following.[156]

Purpose-Based Routing Protocols

The purpose-based routing protocols are those, which are designed and implemented for achieving a particular objective for data packet routing based on IP. The main examples of purpose-based routing protocols include:

- IGP
- EGP

Behavior-Based Routing Protocols

This category is classified in terms of the behavior of the routing protocol. The protocols in this category can support certain behaviors to improve the routing efficiency and reduce the latency in the routing of the information. The example of behavior-based routing protocols include:[157]

- Traditional Classful Routing Protocols
- Classless Routing Protocols

The classful routing protocols are those protocols that send periodic updates of routing tables and do not support the Variable-Length Sub-net Mask (VLSM) and Class-less Inter-Domain Routing (CIDR). These routing protocols are able to route the variable subnet masks for providing detailed information about the connected interfaces. By providing the details of variable-level sub-net masks, the burden of routing the packet to the desired destination is reduced. These are called traditional routing protocols, which use relatively larger bandwidth and put a heavy load on the performance and efficiency of the networks.

The most important examples of classful routing protocols that qualify for this category of routing protocols include the following:

- Routing Information Protocol Version 1 (RIPv1)
- Interior Gateway Routing Protocol (IGRP)

The classless routing protocols are those that support both VLSM and CIDR routing features, which reduce the burden on the link between the routers and increase the performance, reduce the latency, and increase the efficiency of the routing. These routing protocols use *hello* messaging mechanism to check for the availability and send the routing information when a certain change is noticed. The examples of classful routing protocols are listed here:

- Routing Information Protocol Version 2
- Enhanced Interior Gateway Routing Protocol
- Open Shortest Path First (OSPF)
- Intermediate System to Intermediate System (IS-IS)

Operation-Based Routing Protocols

These are those protocols that create the routing table based on certain operations such as calculations of hops, link congestion, network topology, and other factors. This category can be further divided into three major categories as listed here:

- Distance Vector Routing Protocols
- Link State Routing Protocols
- Path Vector Routing Protocols

Some descriptions with the examples of real-world routing protocols of the aforementioned categories are mentioned in the following subtopics.

Distance Vector Routing Protocols

The distance vector routing protocols are those that use the distance vector routing algorithm for calculating the route information. This is the first category of routing table manipulation, which was used in the initial routing protocols. These protocols are not very efficient and suitable for larger networks. The distance vector category of routing protocols is referred to as the first-generation of routing protocols. These protocols are known for the following characteristics for calculating the routing table and network information.[158]

- Each router calculates the knowledge of all routers in a table.
- These protocols use hops as the core value of the routing table.
- They share the connected routers with only neighbors.
- They send updates after every 30 seconds.

The examples of distance vector routing protocols include:

- RIPv1
- ARPANET protocol

Link State Routing Protocols

In the link state routing protocols, the information about the entire network is learned by the connected routers, from which the shortest path to the destination network is calculated. The periodic updates are not sent in the routing protocols belonging to this category. These are also known as the second-generation protocols, which are more efficient than the distance vector protocols. The use of Dijkstra Algorithm is extensively used for the routing protocols under this category.

The major examples of the routing protocols related to this category are mentioned in the following list:[159]

- OSPF
- IS-IS

Path Vector Routing Protocols

This category of routing protocols uses a different algorithm from link state and distance vector. These routing protocols manipulate the routing table with the direct address of the destination network. The router maintains a complete path information to the destination in the routing table, and the table is updated dynamically. The looped updates are discarded at the receiver of the looped information packet. The example of this category of routing protocols include:[160]

• Border Gateway Protocol (BGP)

In the BGP protocol, the autonomous system is used that helps advertise the reachability of the networks very easily. The autonomous system boundary routers, precisely referred to as ASBR, send the path-vector, which provides detailed information about the reachability of the networks in a much easier and efficient way.

Sample Questions and Answers for Chapter 5

Q1. State the main functions of an SNMP manager.

A1. The main functions of an SNMP manager are listed in the following:
• Sending queries to SNMP agents
• Collecting responses from SNMP agents
• Configuring the supported variables in the agents
• Acknowledging the synchronous events from the SNMP agents
• Monitoring of managed devices remotely
• Visual, graphical, and numerical presentation of network, data, and other information
• Correlating different metrics and information within the entire network elements

Q2. What is *GetNextRequest*?

A2. GetNextRequest – This is almost the same message like GetRequest. But, this message is used for discovering the available data with the SNMP agent and instructing the agent to keep sending the data with respect to certain terms and conditions.

Q3. What is MIME?

A3. Multipurpose Internet Mail Extensions, precisely MIME, is a type of protocol that defines different forms of texts, coding systems, formats of data, and many other features of data, which are required to be transferred through emails over the SMTP and other email protocols. As the name indicates, it is an extension, which works over the base email protocol.
  Primarily, this protocol was designed for the SMTP protocol but other email as well as communication protocols also benefited from the definitions of the data formats,

encoding, and other formats. The data formats by the MIME protocol are extensively used in web-based communication protocols such as HTTP and HTTPS. Nowadays, it is extensively supported by all major email protocols.

Q4. What is POP protocol?

A4. Post office protocol (POP) is one of the important protocols used for pulling emails and managing emails on the server and mailboxes. The most common, the POP protocol is referred to as email pulling protocol between email client and server. This protocol is configured on the email client software such as MS Outlook, Thunderbird, and others. Those clients use the POP protocol to communicate with the email servers for accessing emails and managing the emails in the mailbox.

Q5. Define *classful* routing protocol. What are the core characteristics of such protocols?

A5. The classful routing protocols are those protocols that send periodic updates of routing tables and do not support the Variable-Length Sub-net Mask (VLSM) and Class-less Inter-Domain Routing (CIDR). These routing protocols are able to route the variable sub-net masks for providing detailed information about the connected interfaces. By providing the details of variable-level sub-net masks, the burden of routing the packet to the desired destination is reduced. These are called traditional routing protocols, which use relatively larger bandwidth and put a heavy load on the performance and efficiency of the networks.

Wireless and Cellular Communication Protocols

6

WIRELESS COMMUNICATION PROTOCOLS

A set of rules used by the sender and receiver devices of the wireless communication systems to understand and recover the data transported over the wireless channel such as air or vacuum is known as the wireless telecommunication protocol.

Wireless communication has come a long way for over 12 decades now. The first communication protocol developed by Marconi was a simple electromagnetic waveform to be generated and transmitted by a radio transmitter and received by a radio receiver.

DOI: 10.1201/9781003300908-8

The communication of radio signals (analog) signals was much simpler and basic than present-day protocols used in modern wireless communication. The older wireless protocols were mostly dependent on the electronic circuits to transmit the data over the physical layer of the OSI model. The protocols or the communication rules dealing with the higher layers like data link, network, transport, and application layers were not introduced at the early stages. The present-day wireless communication protocols cover all of the layers of the OSI model of telecommunication.

Significant advancements and improvements have been made already in the older protocols dealing with the physical layer of communication. Numerous other protocols dealing with the upper layers of the OSI model of communication have also been developed for establishing the most sophisticated communication in modern telecom systems. The modern wireless communication/network protocols can be divided into two major categories at this point in time:

- Fixed wireless network protocols
- Cellular wireless network protocols

Here, let us talk about some of the most common wireless communication protocols used in modern communication systems powered by modern networking technologies.

FIXED WIRELESS NETWORK PROTOCOLS

Fixed wireless communication protocols are those ones, which do not support the mobility and soft handover or handoff within two separate cells or domains of wireless communication protocols. These wireless protocols are mostly used by the fixed devices located within the periphery of a wireless signal range. The use of these protocols is very high in the office, home, or any other public place. The most important use of fixed wireless protocols is the modern IoT. A few very important fixed wireless communication protocols are mentioned in the following subtopics.[161]

IEEE 802.11 Protocols

Institute of Electrical and Electronics Engineers (IEEE) developed a set of protocols under the 802.11 project, which is an important part of the IEEE 802 project started by IEEE. Numerous sub-protocols based on IEEE 802.11 WLAN protocol have been specified and launched for different wireless services in modern wireless communication. A working group to work on the Wireless LAN protocol was constituted in 1990. With the help of that group, the first version of IEEE 802.11 was launched/released in 1997.[162]

This family of wireless protocols deals with the local area networking devices and Internet access over the radio wireless signals. The data transmission between the wireless devices takes place through the frequency bands such as 2.4, 5, 6, and 60 GHz. The most commonly used frequency for the Wi-Fi compatible devices in IEEE 802.11 specifications is 2.4 GHz. The main features and characteristics of the IEEE 802.11 family of wireless protocols are listed here:

- This family of WLAN protocols consists of numerous specific protocols. A few of those protocols are mentioned here:[163]
  - **IEEE 802.11b** – This protocol uses Direct Sequence Spread spectrum (DSSS) modulation at the physical layer and was released in 1999.
  - **IEEE 802.11a/j/p/y** – All these different versions of WLAN protocols use the OFDM modulation at the physical layer of the OSI model and were released in different years between 1999 and 2010.
  - **IEEE 802.11g** – This WLAN protocol specification defines the use of ERP-OFDM modulation at the physical layer over 2.4 GHz frequency. It was released in 2003.
  - **IEEE 802.11n/ac/ax** – All these three versions of WLAN protocol support MIMO-OFDM modulation over 2.4/5, 5, and 2.4/5/6 GHz, respectively. These versions were released in 2009, 2013, and 2021, respectively.
  - **IEEE 802.11ad/aj** – These both standards support OFDM single carrier modulation. They were released in 2012 and 2018, respectively. They use 60 and 45/60 GHz, respectively.
  - **IEEE 802.11ay** – This standard operates in the 60-GHz frequency band and supports OFDM single carrier modulation. It was released in 2021 and is capable of handling a data rate of 20 Gbps within a range of 100 meters.
- This family of protocols supports half-duplex communication mode through a wide range of over-the-air modulation techniques as mentioned in the afore-mentioned point.
- Uses the transmission control technique known as Carrier Sense Multiple Access with Collision Avoidance for smooth resource sharing in the connected environment of the networking devices within a WLAN range.
- The compatibility and interoperability are managed and governed by the WI-FI Alliance, which is a non-profit group of a large number of electronic and tele-communication companies (over 800) operating across the world.
- This family of protocols supports different variants of DSSS and OFDM signaling/modulation techniques.
- For easy understanding, WI-FI Alliance categorized the protocols of the IEEE 802.11 family into six generations in 2018. The generations are classified as follows:
  - **First-generation WLAN protocols** – IEEE 802.11b
  - **Second-generation WLAN protocol** – IEEE 802.11a
  - **Third-generation WLAN protocol** – IEEE 802.11g
  - **Fourth-generation WLAN protocol** – IEEE 802.11n
  - **Fifth-generation WLAN protocol** – IEEE 802.11ac
  - **Sixth-generation WLAN protocol** – IEEE 802.11ax
- This entire family of protocols is contention-based MAC protocols, which use as low power as possible for efficient transfer of data within a wireless network.

- The communication mechanism of WLAN 802.11 protocols is based on collision avoidance. It uses the following messages for establishing a smooth communication within a WLAN:[164]
  - **Request to Send (RTS)** – This message is used to reduce the collisions within the network to share the wireless resources efficiently.
  - **Clear to Send (CTS)** – This message is generated against the RTS message generated by the sender. The receiver sends this message to inform that it is clear to send the message now.
  - **Acknowledgment (ACK)** – This message is used for sending the confirmation of the received data packet intact without any fault or compromise.

Bluetooth Protocol

Bluetooth Protocol is another member of the IEEE 802 project or family of telecommunication protocol. It works in a type of ad hoc wireless network that communicates in the peer-to-peer mode of networking. This protocol is referred to as IEEE 802.15 protocol specification. The name has been taken to commemorate the King Viking, who unified Norway and Denmark back in the 10th century. The main player behind the idea of Bluetooth technology was Ericsson. It conceived the idea back in 1994 to develop an ad hoc networking protocol for wireless communication in the Industrial, Scientific, and Medical (ISM) band. Later on, other companies joined hands with Ericsson in 1998 to form a Special Interest Group (SIG) for materializing this idea with great efficiency and performance.

The main features and characteristics of Bluetooth wireless protocol are mentioned in the following list:[165]

- It uses FHSS and DSSS modulation technologies.
- Each piconet formed through Bluetooth can support up to seven data channels and three voice channels.
- The maximum range is about 10 meters.
- Supports data speed up to 721 kbps.
- One single piconet can accommodate up to eight devices.
- It consumes about 8–30 milliamperes of current while transmitting data.
- It uses the scattered ad hoc network topology as the core architecture of the connectivity.
- Bluetooth protocol allows the enabled device to co-exist in multiple piconets powered by the Bluetooth technologies.
- Bluetooth architecture defines four states of the devices as follows:
  - Master Station (M)
  - Slave Station (S)
  - Standby Station (SB)
  - Hold/Parked Station (P).
- Bluetooth architecture defines the following protocol stack:
  - Radio Frequency (RF) layer
  - Link Management Protocol (LMP) layer
  - Logical Link Control and Adaptation Protocol (L2CAP) layer

- Service Discovery Protocol (SDP) layer
- Telephony Control Protocol (TCP) layer
- RFCOMM Layer
- Application Layer.
- Bluetooth uses FHSS-CDMA/TDD mechanism for media control within the piconet.
- Operates in 2.4-GHz ISM frequency band.

Bluetooth Low-Energy (BLE) Protocol

Bluetooth Low Energy (BLE) is an efficient version of Bluetooth protocol with additional features to reduce the consumption of energy. It is commonly referred to as BLE protocol or Bluetooth Smart Protocol. Nokia Inc. designed this protocol for strengthening the efficient use of IoT devices in a Bluetooth network environment in highly optimized conditions for power consumption. Presently, this protocol is implemented and governed by Bluetooth SIG (GIS). The main features, characteristics, and capabilities of BLE protocol are listed here:[166]

- Very low energy consuming protocol, which consumes about half the energy consumed by the traditional Bluetooth enabled devices.
- Supports four major types of functions or roles of a device as listed here:
  - Broadcaster function/role
  - Observer function/role
  - Central function/role
  - Peripheral function/role.
- It is industry-specific technology extensively used in the IoT and other domains.

ZigBee Protocol

ZigBee is a very popular wireless protocol, which is application-specific and extensively used for home automation, industrial automation, smart equipment monitoring, and other such types of solutions. It is based on the physical and MAC layer protocol defined under the specifications referred to as IEEE 802.15.4, which is a member of the IEEE 802 family of protocols. The interoperability and compatibility features of this protocol for a wide range of devices are monitored by the ZigBee Alliance like the Wi-Fi Alliance.

It is a machine-to-machine communication protocol, which uses very low power supported by the batteries, which provide a long life of operations. The main features and characteristics of the ZigBee protocol are mentioned here:[167]

- This protocol is built on MAC and physical layer.
- Natively, it is a wireless mesh networking protocol, but also supports tree and star topologies.
- Operates in ISM band with 2.4 GHz frequency, but in certain countries, a few other frequencies are also used for purpose-specific applications.

- The initial idea for this protocol was conceived in 1990s, developed in 2003, and finally released as Zigbee Specifications 2004 in 2005. The revised version of the Zigbee was announced in 2006 that supersedes the previous version released in the name of 2004 specifications.
- It is a wireless personal area network (WPAN) technology, which is capable of transmitting signals between 10 and 100 meters in the line of sight.
- Supports AES 128-bit symmetric encryption key.
- Supports 250 kbps data rate.
- Zigbee defines three types of device operating types as listed here:
  - ZigBee Coordinator (ZC)
  - ZigBee Router (ZR)
  - ZigBee End-Device (ZED).
- Zigbee protocol supports both non-beacon-enabled and beacon-enabled networks simultaneously.
- It uses the channel access mode known as carrier-sense multiple access with collision avoidance CSMA/CA technique.
- The IoT is the most suitable field for this wireless protocol.

6LoWPAN Protocol

The 6LoWPAN is a set of wireless network specifications specifically designed for the IoT applications powered by the IPv6. It uses the low power devices that are run by batteries.[168]

This is an architecture that uses the wireless protocol defined under IEEE 802.15.4. The architecture design is a network of small islands of embedded systems powered by the low-power and low-latency communication like Zigbee and others. Hence, the main objective of this specification is to assist the communication between small islands of embedded networks. The architecture of the 6LoWPAN specification consists of small stub networks powered by the IPv6 protocol and routing protocols that help transmission and communication of the data and signaling.

Near-Field Communication (NFC) Protocol Suite

NFC is a wireless communication technology based on standards and built over the RFID, which uses electromagnetic induction for producing communication signals between the two supported devices. It is a very short-range wireless communication standard, which supports communication within the range of 4 cm (max). This technology is also known as close proximity communication standard. It is commonly used for contactless payment transactions. This technology is also useful in connecting two devices when they come in close contact and also accessing the digital content on the NFC-enabled device.

The main features and characteristics of the NFC wireless protocol suite are mentioned here:[169]

- NFC communication uses a 13.56 MHz frequency for generating electromagnetic waves.
- It supports a low data rate of about 424 bps.

- Supports three modes of communication:
  - NFC Card Emulation
  - NFC Reader/Writer
  - NFC Peer-to-Peer.
- It uses the ISO/IEC 18000–3 air interface.
- The major use cases of NFC communication standard include the following:
  - Contactless mobile payment
  - Access control systems
  - Peer to peer data exchange
  - Ticket purchasing.
- Supports two types of devices as listed here:
  - Polling device
  - Listening device.
- It supports the following modulation and encoding schemes at the physical layer level:
  - ASK
  - BPSK
  - Non-Return-to-Zero-Level
  - Non-Return-to-Zero-Inverter
  - Manchester encoding
  - Modified Miller encoding.

CELLULAR MOBILE COMMUNICATION PROTOCOLS

Cellular mobile communication protocols are those different sets of rules that realize the cellular communication among the mobile devices satisfying all the features and capabilities of the cellular communication network. Cellular mobile technologies for the last-mile air interface are changing drastically for the increased capacity and bandwidth to cater to the increased demands from the customers. With changes in the technology, the changes in the protocols are bound to happen, especially at the physical and data link layers. Those changes are governed by the new wireless protocols used in cellular mobile networks.

Let us talk about a few most common protocols that work at the crust of the entire core network that a cellular mobile network uses. Various details of the wireless technologies and related protocols and encoding schemes have already been discussed previously in various parts.

CAMEL Architecture of WIN Services

Before diving into the CAP protocol, we should have a look at the Customized Applications for Mobile-networks Enhanced Logic (CAMEL) application stack, also known as service stack based on the IN. CAMEL is a service architecture, which was designed for developing different types of features, capabilities, and controls in the cellular mobile services based on Global Mobile System (GSM) and UMTS.

CAMEL architecture was standardized and defined by the ETSI specifications. The CAMEL architecture was released in different phases with additional capabilities and backward compatibilities. For example, the first phase was released in 1996–97. This version only supported a few basic call functionalities, features, and controls. The second phase was released in 1997–98 shortly after the first release. This logic of service was more capable than the previous one. It was backward compatible in terms of services released in the first phase. It had many additional features and capabilities of the services such as more event detection points, call deflection, explicit call transfer, multi-party calls, and many other services.[170]

The third phase was released in 1999 as a part of the 3rd Generation Partnership Project (3GPP). This release was capable of handling roaming features, GPRS sessions, dialed services, and many others. The fourth version or phase of the CAMEL architecture was released in 2002. This release supported all the previous features, capabilities, and call controls along with many new features and capabilities. A few of them include adding more calls to existing call party, support for messages in both packet and circuit-switched networks, IP multimedia subsystem, different tones, and many other features.

The CAMEL architecture deals with the four entities in mobile cellular networks. These entities are logical entities residing on certain physical entities as mentioned here:

- **GSM Service Switching Function (GSM-SSF)** – This logical entity resides over the MSC (Mobile Switching Center) of the cellular/mobile network. The main function of this function is to switch the call from one route to another route in the GSM network. Real-time billing is handled by this FE and, finally, dispatched to the SCF function.
- **GSM Specialized Resource Function (GSM-SRF)** – This logic entity deals with specialized resources such as voice commands, announcements, pin, and number digit collection. This function logically resides as a control function on SCP along with the interface with the specialized resources such as Automated Voice Record and other such types of resources used in the network.
- **GSM Service Control Function (GSM-SCF)** – This logical module of service architecture resides in SCP. This functionality deals with the control of the services such as billing, roaming, value-added subscription, and much more. This function is the fundamental part of the Mobile Wireless IN (WIN).
- **GPRS Service Switching Function (GPRS-SSF)** – This function is similar to the GSM-SSF located on the MSC. It is the data service node for switching the data services to the mobile user in communication with the main SCF.

The communication among the aforementioned functionalities for control and providing value-added services is done through the CAP protocol, which is described in the following topic with full details and designs.

CAP Protocol

CAP is an intelligent service handling protocol for CAMEL services in the cellular mobile networks based on GSM (Global System for Mobile Communications), UMTS, and GPRS (General Packet Radio Service) technologies. CAMEL is a stack of IN services such as prepaid cards, number portability, call transfer, and many other similar kinds of services defined and implemented on the IN node known as SCP. The CAP protocol is used to communicate between the mobile switching center (MSC) and SCP.[171]

The CAP protocol is like the INAP protocol used in the fixed telephone systems for realizing the value-added services over the traditional landlines. This protocol is extensively useful for the operators who offer mobile cellular services based on different technologies simultaneously, which are commonly referred to as Public Land Mobile Network (PLMN).

The CAP protocol supports the TCAP protocol on the SS7 signaling system and works above the TCAP protocol. It is also known as the ROSE user protocol. The realization of the services in CAP protocol is developed upon the previously adopted fixed telephony network protocols for the value-added services like toll-free numbers, prefix-less dialing, universal numbers, and many other similar kinds of services.

The main features and characteristics of the CAMEL Application Protocol (CAP) are mentioned in the following list:

- A popular signaling communication protocol extensively used for the IN services in the wireless networks consisting of GSM, 3G, and beyond.
- Extensively useful for Home PLMNs and even beyond that in the cellular mobile networks.
- CAP protocol was also revised four times with the advent of the enhanced phase of the CAMEL service architecture as mentioned in the aforementioned topic.
- CAP protocol is defined and specified in terms of service aspects and technical realization. Both of those specifications are standardized in 3GPP TS 22.078 and 3GPP TS 23.078 standards.
- The CAP protocol can be defined with the help of three different aspects or sections as mentioned here:
  - The operations between different entities are defined by the use of ASN.1.
  - State transition diagrams define the actions carried out at each entity in the wireless IN network of CAP protocol.
  - Two kinds of associations are defined in the specifications of CAP protocol they are known as SACF and MACF.
- Supports QoS capability through choosing the classes either class 0 or class 1 of the SCCP protocol stack.
- The role of CAMEL protocol becomes very important in case a mobile subscriber is in the roaming state. CAMEL has to monitor and control the roaming services.
- CAMEL protocol is based on the following entities in the GSM network:
  - GSM Service Control Function (gsmSCF)

- GSM Service Switching Function (gsmSSF)
- GSM Specialized Resources Function (gsmSRF)
- GPRS Service Switching Function (gprsSSF).

With the help of CAMEL protocol, a cellular network is able to deliver a wide range of intelligent services and features to the end-users in the cellular mobile networks. The scope of wireless INs is huge in the future as the new types of value-added services, features, and capabilities will continuously be added to modern wireless technologies.

Sample Questions and Answers for Chapter 6

Q1. What are the two major categories of wireless communication/network protocols?

A1. The modern wireless communication/network protocols can be divided into two major categories:
- Fixed wireless network protocols
- Cellular wireless network protocols

Q2. Name the six generations of IEEE 802.11 family protocols according to the WI-FI Alliance.

A2. WI-FI Alliance categorized the protocols of the IEEE 802.11 family into six generations in 2018. The generations are classified as mentioned here:
- **First-generation WLAN protocols** – IEEE 802.11b
- **Second-generation WLAN protocol** – IEEE 802.11a
- **Third-generation WLAN protocol** – IEEE 802.11g
- **Fourth-generation WLAN protocol** – IEEE 802.11n
- **Fifth-generation WLAN protocol** – IEEE 802.11ac
- **Sixth-generation WLAN protocol** – IEEE 802.11ax

Q3. What is an RTS message?

A3. Request to Send (RTS) – This message is used to reduce the collisions within the network to share the wireless resources efficiently.

Q4. What is Bluetooth?

A4. Bluetooth Protocol is another member of the IEEE 802 project or family of telecommunication protocol. It works in a type of ad hoc wireless network that communicates in a peer-to-peer mode of networking. This protocol is referred to as IEEE 802.15

protocol specification. The name has been taken to commemorate the King Viking, who unified Norway and Denmark back in the 10th century. The main player behind the idea of Bluetooth technology was Ericsson. It conceived the idea back in 1994 to develop an ad hoc networking protocol for wireless communication in the Industrial, Scientific, and Medical (ISM) band. Later on, other companies joined hands with Ericsson in 1998 to form a Special Interest Group (SIG) for materializing this idea with great efficiency and performance.

Q5. What is the frequency used by NFC communication for generating electromagnetic waves?

A5. NFC communication uses a 13.56 MHz frequency for generating electromagnetic waves.

Fiber Optic Data Transmission Protocols

7

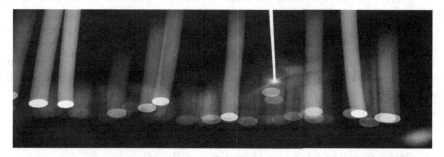

Fiber optic transmission has become very popular in modern telecommunication systems due to its higher capacity and no electromagnetic interference. Fiber optic is being used in all types of networks such as metro fiber networks, last-mile networks, backbone networks, inter-continent networks, and others. Numerous types of fiber optic transmissions and multiplexing protocols have got deeper traction in the telecommunication sector.

In this chapter, we will talk about a few very important fiber optic transmission/communication protocols extensively used in the modern telecom sector.

WAVELENGTH DIVISION MULTIPLEXING (WDM)

Wavelength division multiplexing, precisely referred to as WDM, is a type of multiplexing of multiple signals of different wavelengths into one composite signal for transmission over a long-haul fiber optic network.

The fiber optic uses the light signal of a certain frequency generated by the laser diode to transmit into the fiber optic core for traveling to the receiver end without getting interfered with electromagnetic waves. As the signal sent out over the fiber optic cable is the light signal, it is not affected by any electromagnetic interference. This is the simple

DOI: 10.1201/9781003300908-9

FIGURE 7.1   Wavelength division multiplexing.

form of fiber optic transmission. But, sending multiple signals over the same fiber optic is possible through the WDM. The schematic diagram of WDM is shown in Figure 7.1.

In WDM, the light signals of different wavelengths are combined through a multiplexer, which receives the signals from different transponders and combines them in terms of their wavelengths and sends them out on the signal fiber cable. That signal is demultiplexed at the receiving end by using a splitter or demultiplexer, commonly known as DEMUX at the receiving end. This is how the signals are sent to the respective receiver at the far end. This process is known as demultiplexing of wavelengths.[172]

Wavelength multiplexing was introduced in 1970s, but at that time, the cost of fiber optic was so high and the technology could only multiplex two wavelengths at a time. However, with the course of time, more improvements started materializing, and at this time of writing this book, over 160 channels can be multiplexed simultaneously.

The WDM is further divided into two categories:

- Coarse Wavelength Division Multiplexing (CWDM)
- Dense Wavelength Division Multiplexing (DWDM)

Coarse Wavelength Division Multiplexing (CWDM)

The CWDM is a type of WDM, which has a low-capacity and low-power consuming mechanism. It can multiplex eight light channels of different wavelengths. The spacing between the two wavelengths is 20 nm. The supported distance is much lesser than the other type of WDM known as DWDM. This multiplexing operates in the Erbium Window, which is also known as the C-band of optical frequencies. The C-band wavelength used in the CWDM multiplexing is 1,550 nm.

Dense Wavelength Division Multiplexing (DWDM)

DWDM is the advanced version of the WDM. It can support more channels of different wavelengths to multiplex and send over the single-mode fiber optic cable to the far end. The main features and characteristics of this multiplexing technique are listed here:[173]

- Operates in the C-band as well as in the L-band.
- It can multiplex 40 and 80 channels in the C-band with wavelength 1,550 nm.

- In the L-band (1,565–1,625 nm), it can support up to 160 channels to double the capacity of the multiplexing module.
- Uses inter-channel spacing of 50 GHz for 80 channels and 100 GHz for the 40-channel multiplexing system.

FREE-SPACE OPTICAL NETWORK (FON)

Free-space optical network, precisely referred to as FON, is a new optical technology powered by the infrared (IR) light signals to be transmitted over the atmosphere or space for long-haul optical communication. The major mechanism or protocol for the transmission of the optical signal is almost resembling the same fiber optic transmission with just a few modifications.

The main features, capabilities, and characteristics of this optical communication model are listed here:[174]

- It uses near-IR visible wavelengths for optical transmission.
- It consumes very reduced power about half of the RF-based transmission.
- Supports a higher level of data security.
- The reduced antenna size is almost ten times lesser than the RF antenna.
- Supports a large number of architectures for both indoor and outdoor optical transmission.
- Supports huge wavelength range between 700 and 1,600 nm.
- Able to support over 100 THz bandwidth, which is about five times the bandwidth of RF frequency transmission into the atmosphere.

STANDARD G.651

Standard G.651 is an ITU recommendation for the use of multimode fiber optic cable. This recommendation is extensively used for the fiber optic cabling in the multi-tenant building for IT services, telecom services, or even for the home and office cabling systems. This recommendation has been revised with the G.651.1 recommendation, which is in force now. The main characteristics of this recommendation are listed here: [175]

- Multimode fiber cable of 50/125 μm graded index is recommended for the premises cabling as well as for the IT/Telecom cabling, which has been mostly replaced by the single-mode fiber cabling in the industry. The 50 indicates the core of the fiber optic and 125 μm defines the cladding of the multimode fiber optics.
- Supports the system bit rate of above 10 Gbps.
- Suitable for the use in the range of either 850 or 1,300 nm bands.

STANDARD G.652

This is an international standard for one of the single-mode fiber optic cables defined by the ITU. ITU has standardized six types of single-mode fiber optics under the recommendations known as G.652, G.653, G.654, G.655, G.656, and G.657. The entire series of recommendations is known as the G.65x family of recommendation. In this family of standards, two fibers optics defined under G.652 and G.655 are extensively used in the telecommunication and IT sectors. The other versions are used very rarely. [176]

The main features and characteristics of G.652 fiber optic standards are mentioned in the following list:

- Alternatively, it is known as standard SMF/zero depression-shifted fiber.
- Operates in 1,310, 1,550, and 1,625 nm wavelengths.
- Classified into four major categories referred to as A, B, C, and D categories.
- Category G.652A and B are used in the 1,310 nm range.
- Category G.652C and D are used in the higher wavelength bands.
- Cat C and D are suitable for CWDM applications.
- For the DWDM applications, the three variants of G.655 single-mode fiber optics are used.

STANDARD G.983

It is an international standard applicable for narrowband and broadband services over the passive Fiber Optic Network commonly referred to as FON. This standard has been extended with the following series of recommendations for the additional services:[177]

- **G.983.1** – For Broadband Passive Optical Network (B-PON)
- **G.983.2** – This standard specifies the control and management of Optical Network Terminal of B-PON
- **G.983.3** – For the broadband optical services and wavelength allocation
- **G.983.4** – For broadband services through dynamic bandwidth assignment
- **G.983.5** – For broadband with improved survivability

All of the aforementioned components specify the efficient, effective, and standard narrowband as well as broadband services over the passive optical network. These recommendations cover the fiber-related issues that provide broadband services based on the ATM network over the passive optical fiber network.

STANDARD G.984

This standard specifies the distributed service network at the last-mile network referred to as Gigabit Passive Optical Network, precisely as GPON. This standard is developed by the ITU for the implementation of fiber to the premises (FTTP) concept of a fiber-optic-powered last-mile network. Till today, this standard has been revised and released seven times with different versions of this standard.[178]

The main features and characteristics of G.984 recommendations are listed here:

- Supports uplink 1.2 Gbps and downlink 2.4 Gbps
- Supports up to 128 ONTs
- Uses Reed-Solomon algorithm for error correction
- The first version of GPON standard was ratified in 2003

PLESIOCHRONOUS DIGITAL HIERARCHY (PDH)

PDH is a technology used for the transmission of bulk volumes of data streams over optical and microwave transmission. This technology uses the combinations of data links in a hierarchical order to make bigger bunches. For example, a stream of 2.048 Mbps is used as the fundamental link, which makes different bunches of data like STM 1, STM 4, and others. This technology has now been replaced by the fully synchronous and digital transmission known as SDH.[179]

SYNCHRONOUS DIGITAL HIERARCHY (SDH)

This is another important transmission technology that supports the merger of small data streams into a synchronized container to make it a bulk volume of data into different large-scale streams of data transported over a long-haul backbone network based on fiber optics. This standard technology for transmission also supports microwave and traditional interfaces of data transmission.

The SDH technology is commonly used in Europe, while its US counterpart is known as the SONET (discussed later separately), which uses the same SDH technology but with different bit rates for input and output interfaces. The main features and characteristics of SDH are listed here:[180]

- It was first developed in 1985 to replace the near-synchronous transmission technology known as PDH, which is mentioned in the aforementioned topic.

- It is backward compatible technology to implement PDH streams too.
- It is standardized by ITU.
- It supports the transmission of ATM, ISDN, Ethernet, SAN, and PDH signals simultaneously.
- This uses a synchronous clock system for combining the "S" number of streams of "D" rate of bitstreams into the "S x D" container on a synchronously clocked network.
- It supports both the low-order and high-order multiplexing by using the function of add-drop multiplexer, commonly known as ADMs.
- It supports multiple high-order multiplexed streams like STM1, STM 4, STM8, STM 16, etc.
- At low order, it supports E1 and sub-connections of a wide range of data streams.
- It defines three types of interfaces – one for network node and two for the user-network interfaces as listed here:
  - Network Node Interface (NNI)
  - User Network Interface for customer
  - U interface for Broadband-ISDN support.
- It uses the dedicated communication channels within the interface rate for management functionalities.
- SDH consists of ADMs, Hubs, Digital Cross-Connects, and different interfaces.
- It supports ring, star, mesh, linear, and cascading network topologies.
- The frame of the SDH container consists of nine equal segments shared within a period of 125 microseconds.
- The size of the SDH container is a multiple of 9 rows and 270 columns. Each byte of nine slots consists of 64 kbps. Thus, the total data rate will be 576 kbps.
- The first nine columns contain the section overhead information and additional features or management functions.
- This technology is also standardized in the G.709 recommendation.

The virtual container of the SDH technology contains numerous tributary units referred to as TUs, tributary unit groups (TUGs), administrative units (AUs), and virtual containers (VCs). They are named with number scheme such as 1, 2, 3, and so on. These units also work as a pointer for the data in the virtual container units.

SYNCHRONOUS OPTICAL NETWORKING (SONET)

Synchronous optical networking, precisely referred to as SONET is an optical networking transmission protocol for multiplexing small bit streams into large-scale transmission streams over the optical network at the backbone transmission network. This protocol is developed by ANSI. This protocol is designed to govern the transmission of data streams over optical fiber networks. It defines different standards for the transmission line rates.[181]

The main characteristics and features of SONET are mentioned here:

- Supports the base rate of 51.84 megabits per second, which is also known as Synchronous Transport Signal or STS.
- Extensively used in North American countries like the USA, Canada, and others.
- The STS is also known as Optical Carrier or OC with different levels.
- The data speeds of the connections are specified as under:
  - **STS-1/OC-1** = 51.84 Mbps
  - **STS-12/OC-12** = 622.08 Mbps
  - **STS-96/OC-96** = 4976.64 Mbps.
- All components and protocols supported by the SONET protocol are almost the same as those supported by the SDH protocol. The main difference is the data rates. The data rates in both protocols are different for every interface, stream, management channel, tributaries, administrator units, and others.

OPTICAL TRANSPORT NETWORK (OTN)

Optical Transport Network, precisely OTN, is a new standard for transporting the data based on WDM, Wavelength Switched Optical Network (WSON), and TDM devices simultaneously and transparently over a single optical interface. This standard is specified through G.709 recommendation, which is developed to supersede the existing SONET/ SDH systems and replace them with advanced technology which will offer more transparency, efficiency, performance, and many other benefits.

This architecture is based on multiple layers that can accommodate different types of data transportation as mentioned earlier. Those layers, commonly referred to as sub-layers, are responsible for certain specific functions and services. Those sub-layers are normally activated at their respective termination points. This system defines the hierarchy of Optical Data Units (ODUs) such as ODU1, ODU2, ODU3, ODU4, and others. These data rates of those ODUs are different. The ODU1s can be contained into the ODU2s and ODU2s can be housed into the ODU3s and so on. Meanwhile, the different ODUs like ODU1 and ODU2 can be multiplexed into the ODU3. This capability makes this technology so efficient and promising for future needs.

The main features and characteristics of the OTN standard are mentioned here:[182]

- ODUs define five basic bit rates for client signals, i.e., 1.25, 2.5, 10, 40, and 100 Gbps by containing different levels of ODUs.
- The layered frame formats are shown in Figure 7.2.
- Supports WDM, TDM, and WSON equipment and devices simultaneously.
- It supports two types of switching layers WDM and WSON.
- Numerous functions related to data transportation such as routing, survivability, multiplexing, management, transport, and supervision are specified in this architecture/standard.
- OTN defines six distinct frame formats as mentioned here:

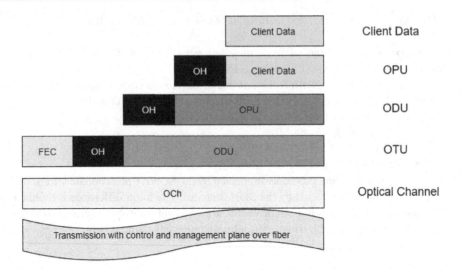

FIGURE 7.2 Wrapping of overheads into frames (here, FEC means "Forward Error Correction", while OH stands for "OverHead").

- **Optical-Channel Payload Unit (OPU)** – This frame contains client data and overhead defining the type of data.
- **ODU** – This frame format is further divided into sub-layers defined as ODU1, ODU2, and so on.
- **Optical Transport Unit (OTU)** – This is also known as a physical port with monitoring and FEC headers.
- **Optical Channel (OCh)** – It is an end-to-end optical connection or path between two supported devices.
- **Optical Multiplex Section (OMS)** – This layer governs the DWDM between two Optical Add-Drop Multiplexers (OADM).
- **Optical Transport Section (OTS)** – This layer governs the DWDM multiplexing between two relays.
- The control plane of the OTN supports the Generalized Multiprotocol Label Switching, precisely known as GMPLS for controlling, management, and other networking maintenance functions in the architecture.

FIBER DISTRIBUTED DATA INTERFACE (FDDI)

Fiber Distributed Data Interface, commonly referred to as the FDDI interface standard in the telecommunication sector, is an ANSI standard for self-healing LAN fiber connectivity.

This standard has also been adopted by the ISO for international implementation. This standard defines the network interface connectivity on the basis of Media Control Access (MAC) sub-layer, which monitors the dual fiber cables used in this interface.

The one fiber cable is active and takes the traffic, and the other one is in hot-active standby mode. If the main fiber optic cable is disturbed anywhere, the traffic/data is shifted to the standby cable. Currently, FDDI is mostly used in a ring configuration, which allows all devices to remain connected in case of a single fiber break.

The main features and characteristics of the FDDI are mentioned here:[183]

- The physical medium of the FDDI standard interface is fiber optic.
- It supports both single-mode and multimode fiber optic cables.
- It is based on the Physical and MAC layers.
- It supports as much as 100 Gbps data over a distance of about 200 km.
- It has dual rings of fiber connected for the reliability and redundancy.
- It uses the mechanism of token passing defined by the IEEE 802.4 token bus recommendation.
- The field of an FDDI frame includes the following:
  - **Preamble** – 1 byte for synchronization
  - **Start delimiter** – 1 byte to specify the beginning of the frame
  - **Frame control** – 1 byte to specify the data type
  - **Source address** – 2–6 bytes
  - **Destination address** – 2–6 bytes
  - **Payload** – Variable size
  - **Checksum** – 4 bytes for detecting error and frame sequence
  - **End delimiter** – 1 byte to specify the end of the frame.
- This standard is extensively used in the backhaul as well as distribution networks powered by the fiber optic services, especially in the metro telecom services based on FTTP and related services.

Sample Questions and Answers for Chapter 7

Q1. What is CWDM?

A1. CWDM stands for "Coarse Wavelength Division Multiplexing", which is a type of wavelength division multiplexing with a low-capacity and low-power consuming mechanism. It can multiplex eight light channels of different wavelengths. The spacing between the two wavelengths is 20 nm.

Q2. Define DWDM.

A2. DWDM (Dense Wavelength Division Multiplexing) is the advanced version of the wavelength division multiplexing. It can support more channels of different wavelengths to multiplex and send over the single-mode fiber optic cable to the far-end.

Q3. What are the main features of Free-Space Optical Network (FON)?

A3. The main features of Free-Space Optical Network (FON) are:
- It uses near-infrared visible wavelengths for optical transmission.
- It consumes very reduced power about half of the radio frequency RF-based transmission.
- It supports a higher level of data security.
- The reduced antenna size is almost ten times lesser than the RF antenna.
- It supports a large number of architectures for both indoor and outdoor optical transmission.
- It supports a huge wavelength range between 700 and 1,600 nm.
- It is able to support over 100 THz bandwidth, which is about five times the bandwidth of RF frequency transmission into the atmosphere.

Q4. What is Plesiochronous Digital Hierarchy?

A4. Plesiochronous Digital Hierarchy is a technology used for the transmission of bulk volumes of data streams over optical and microwave transmission. This technology uses the combinations of data links in a hierarchical order to make bigger bunches.

Q5. What are the major data speeds of the connections used in SONET (Synchronous Optical Networking)?

A5. The data speeds of the connections are specified like:
- **STS-1/OC-1** = 51.84 Mbps
- **STS-12/OC-12** = 622.08 Mbps
- **STS-96/OC-96** = 4976.64 Mbps

PART THREE

Digital Security and Privacy Protocols

Network Security Protocols

8

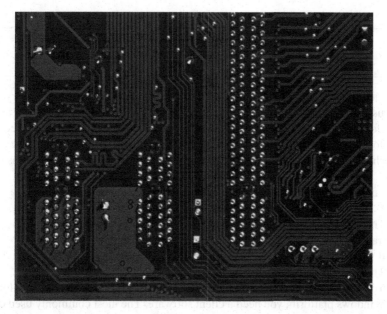

The network security protocols are the standard rules of communication that ensure the integrity, security, and reliability of the data transported over the network connection irrespective of the underlying communication protocols, physical interfaces, platforms, and telecom connection type. The security protocols can deal with a wide range of functions and applications in communication over web interfaces and other communication networks. A few examples of such functions dealt with by the network security include the authorization and authentication of the users, encryption and decryption of the data, safeguarding the integrity of data, and others.[184]

A security protocol prohibits the unauthorized user, service, or software application to access the data or other resources of a network. As the hackers get more learned and employ sophisticated approaches to breach into the data, the security protocols for encrypting the data during the transportation, retrieval, storage, and sending to remote users are implemented nowadays to create more robust and reliable security on the networks and data. A few very important network security protocols are listed here:

DOI: 10.1201/9781003300908-11

- Secure Shell (SSH) Protocol
- HTTPS
- Secure Sockets Layer (SSL) Protocol
- Internet Protocol Security (IPsec) Protocol
- Kerberos Protocol
- Secure Real-Time Transport Protocol (SRTP)
- TLS Protocol
- VPN Protocol

Now, let us discuss a few very important network-security protocols that are commonly used in our modern web and telecom networks.

SECURE SHELL (SSH) PROTOCOL

Secure shell protocol or commonly referred to as SSH protocol is one of the very popular and robust security protocols in the domain. It was developed on the Linux and Unix systems for establishing a secure connection to access the remote host over a secure connection for the remote administration in the networking environment, which uses encryption mechanisms while transporting the data in the shape of commands or files over the network.

The main features and characteristics of SSH are mentioned here:[185]

- This protocol works great for both remote authentication/login and secure transfer of data over the network between two hosts.
- It was first created in 1995 by Tatu Ylönen of Helsinki University Finland.
- This protocol works on the basis of server–client-based communication.
- You can issue direct commands on Linux and Unix platforms and for the Windows platform, you need a client software. The most commonly used client software is PuTTY.
- SSH uses three types of encryption techniques as listed in the following:
  - **Asymmetric encryption** – In this form of encryption, two separate keys are used for encryption and decryption. Those keys are referred to as private key and public key, respectively. Both of those keys work together in the form of a pair, which is named the public–private key pair.
  - **Symmetric encryption** – For this encryption, the secret key, commonly known as shared key or shared secret, is required for encryption as well as for decryption of the data or message at both the client computer and the host machine.
  - **Hashing** – Hashing is a more secure and robust method used by the SSH protocol. This mechanism does not use the keys, but it generates the cryptographic hashed output, which is recovered by the remote host through a cryptographic hash.

- The SSH protocol establishes communication over the TCP port number 22 by default.
- The main uses of SSH protocol include the following:
  - Execution of commands remotely
  - Accessing remote resources securely
  - Delivering software updates and patches
  - Automated and interactive file/data transferring.
- The successful establishment of a connection between two remote machines through SSH protocol is based on two main activities as listed here:
  - Both parties' client/host must agree on the encryption method
  - Client should provide the valid authentication credentials.

The SSH protocol has proven as a very good alternative to insecure Telnet, rLogin, and FTP protocols.

HYPERTEXT TRANSFER PROTOCOL SECURE (HTTPS)

Hypertext Transfer Protocol Secure, precisely known as HTTPS, is the secure version or extension of the basic HTTP protocol for establishing web communication between the web browser and the web server. The HTTP protocol has already been discussed in Chapter 5 at length. This protocol does not use encryption for transporting the requests and responses over the network. The requests and responses are sent out in plain text, which can easily be intercepted and hacked for malicious purposes. To overcome that vulnerability of HTTP, an advanced version that encrypts the requests and responses before sending them out on the network. Thus, the communication over the network becomes very secure and no hackers can breach the encryption of those messages traveling over the network.

HTTPS protocol uses TLS or SSL encryption method. In the HTTPS communication, first of all, the browser verifies the digital certificate of the website that a customer types into the address bar. If the digital certificate is issued from the legitimate authorities, the browser starts communicating with the website through requests and responses in the encrypted messages over the network. HTTPS is indicated with a sign of lock in the address bar as shown in Figure 8.1. It can also be written like www.abcdef.org in the address bar.

The main features and characteristics of the HTTPS protocol are mentioned in the following: [186]

- It is referred to as HTTP over secure connection.
- Uses a valid digital certificate or SSL/TLS certificate issued from the authorized registrars or authorities.

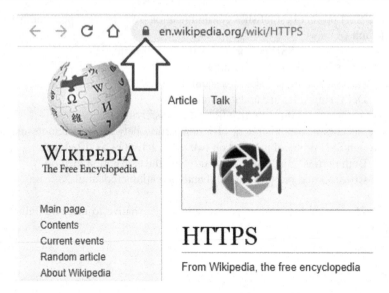

FIGURE 8.1 HTTPS lock sign.

- The responses and requests are sent out over the network as an encrypted code rather than plain text messages in the HTTP protocol.
- Two keys – public and private keys – are used in the communication. The server sends the public key to the client through the SSL certificate. This key is used for generating a session key, which governs the entire communication within that session.

SECURE SOCKET LAYER/TRANSPORT LAYER SECURITY (SSL/TLS) PROTOCOL

SSL is a very powerful and extensively used web security protocol for making sure the integrity, security, and privacy of data and web communication over the Internet. The importance of SSL is very high because, before the implementation of SSL encryption protocol, the messages as well as data would travel in the plain text and in an unencrypted form, which could easily be compromised by any malicious users or hackers.

This protocol was introduced in 1995. Later in 1999, the SSL protocol was succeeded by the TLS protocol with additional advanced features and capabilities. The latest version of the TLS protocol is TLS 1.3, which was released in 2018. The main features and characteristics of SSL/TLS protocol are listed here:[187]

- It offers security to the communication and data between a web browser and a web server.

- It supports the customized application of the SSL protocol suite.
- It was developed by Netscape Inc.
- It uses four major protocols within the entire process of TLS implementation over the HTTP communication as listed here:
  - SSL record protocol
  - Handshake protocol
  - Change-cipher specification protocol
  - Alert protocol.
- Mostly, the aforementioned four protocols work over the TCP/IP suite and are application-specific protocols. It can easily integrate with the HTTP application protocol along with the aforementioned four protocols in secure HTTP communication.

IPsec PROTOCOL

IPsec is the acronym of Internet Protocol Security, precisely IP security. This is a security protocol used for the VPN communication over public networks like the Internet. This protocol is developed and recommended by the IETF. This is a complete suite of security protocols that offer full security, privacy, and data integrity between two endpoints of communication systems. So, it is known as end-to-end security protocol. As the name suggests, this protocol works over the IP network to provide security between two VPN servers, two end-users, VPN-to-client, and client-to-server.[188]

The main features and characteristics of the IPsec protocol are mentioned in the following list:

- This protocol defines numerous communication items such as key management, secure key exchange, encrypted/decrypted packets, authenticated packets, and others.
- The main components of the IPsec protocol are mentioned here:
  - **Encapsulating Security Payload (ESP)** – Encryption, data integrity, authentication for payload, and headers are provided by this component.
  - **Internet Key Exchange (IKE)** – Exchanges keys and finds the way over the security association (SA).
  - **Authentication Header (AH)** – Excepting encryption, the anti-replay, integrity, and authentication are provided by it.
- Other than VPN, this security protocol can be used in the application-level security and IP routing-level security easily.
- It is implemented in two major supported modes of operations as listed here:
  - Tunnel mode of operations
  - Transport mode of operations.
- It supports two types of data packet encodings as mentioned here:
  - **Authentication Header (AH)** – Used for providing the integrity and authentication of data packets

- **Encapsulating Security Payload** – Used for the data confidentiality, data encapsulation, and data encryption

KERBEROS PROTOCOL

Kerberos is one of the most popular Internet security protocols used for the purpose of authentication of valid users to any service or a network host. Microsoft Windows uses it as the default authentication security protocol for its operating systems. This protocol is an open-source security protocol and free to use for all. It is governed by a consortium named Kerberos Consortium. This protocol is extensively used for website user authentication and single sign-on implementation in a wide range of operating platforms like Linux, Apple OS, MAC, FreeBSD, and others. Initially, this protocol was developed by the Massachusetts Institute of Technology under the project named "Project Athena" in the late 1980s.

The main features and characteristics of the Kerberos authentication protocol are as follows:[189]

- The major components of the Kerberos protocol are three depicted through the three heads of dogs in the monogram of the protocol. They include the Client, the Server, and the Key Distribution Center or KDC.
- KDC is used for the third-party authentication services by two major functions known as ticket granting and authentication.
- A KDC component consists of three items as listed here:
  - Authentication Database (DB)
  - Authentication Server (AS)
  - Ticket Granting Server (TGS).
- It supports a wide range of encryption schemes and techniques commonly available in the marketplace.

SECURE REAL-TIME TRANSPORT PROTOCOL (SRTP)

Secure Real-Time Transport Protocol, precisely referred to as SRTP, is an extension protocol of the Real-Time Transport Protocol (RTP), which is used for the real-time transmission of time-sensitive data like VoIP, video stream, and other data. The RTP protocol is less secure and has certain security flaws, which make it prone to security threats in the Internet environment. To overcome those flaws, the secure version of RTP known as SRTP was developed.

It is an ITEF protocol specified under the RFC 3711. In most of the cases, the SRTP protocol is used in conjunction with the other communication protocols rather than used as a standalone protocol in secure communication. Examples of the protocols used in conjunction with the SRTP include SDP, RTSP, and others.

The main characteristics and features of the SRTP protocol are as follows: [190]

- In addition to the RTP features, SRTP has capabilities of message authentication, data confidentiality, replay protection, and the encryption of the communication messages and data between two identified dealing with the real-time communication.
- It uses the robust encryption mechanism known as AES as its default encryption technique.
- It is highly flexible to supports new encryption algorithms if required.
- It uses the latest techniques known as "Salting Keys" for encountering the time-memory tradeoff and pre-computation cyberattacks.
- Frequent refreshing of session keys is also an important feature of the SRTP protocol to strengthen the security of a session in real-time services.
- The SRTP has defined two additional fields in the communication frames other than the RTP protocol, which are used for additional security purposes. These fields are mentioned here:
  - **Master Key Identifier (MKI)** – Used for the key management purposes not for the SRTP cryptographic context.
  - **Authentication Tag (AT)** – This is used for message authentication of data.
- Supports NULL Cipher in case no confidentiality for the data is required.

Sample Questions and Answers for Chapter 8

Q1. What type of communication is the basis of Secure Shell (SSH) protocol?

A1. SSH protocol works on the basis of server–client-based communication.

Q2. What is the meaning of the lock sign for HTTPS address?

A2. It means that the website is secure to browse.

Q3. Name the four major protocols with the entire process of TLS implementation over the HTTP communication.

A3. The four major protocols within the entire process of TLS implementation over the HTTP communication are:
- SSL record protocol
- Handshake protocol

- Change-cipher specification protocol
- Alert protocol

Q4. What does IPsec stand for?

A4. Internet Protocol Security.

Q5. What is Kerberos?

A5. Kerberos is one of the most popular Internet security protocols used for the purpose of authentication of valid users to any service or a network host. Microsoft Windows uses it as the default authentication security protocol for its operating systems. This protocol is an open-source security protocol and free to use for all. It is governed by a consortium named Kerberos Consortium. This protocol is extensively used for website user authentication and single sign-on implementation in a wide range of operating platforms like Linux, Apple OS, MAC, FreeBSD, and others. Initially, this protocol was developed by the Massachusetts Institute of Technology under the project named "Project Athena" in the late 1980s.

Wireless Security Protocols

9

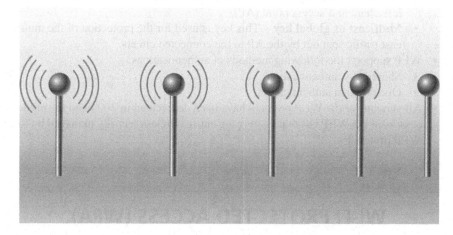

The security of wireless communication is more important than the wired network security due to the wireless network's/system's geographical openness and vulnerability to physical-level signal security. The wireless technology can be used in a wide range of networks such as wide area networks, metro area networks, LANs, and numerous types of micro-sized networks in modern wireless technologies.

In this chapter, we talk about a few very important wireless security protocols used in a wide range of wireless networks used in metros and home/office in modern communication settings.

WIRED EQUIVALENT PRIVACY PROTOCOL (WEP)

Wired Equivalent Privacy Protocol, precisely referred to as WEP protocol, is a type of security protocol used for different types of wireless networks specified under the IEEE 802.11 standard. This security protocol has now been superseded by the Wi-Fi Protected Access (WPA) protocol, which will be introduced shortly after this short introduction (to first understand the necessary technical background).

DOI: 10.1201/9781003300908-12

This security protocol was ratified as the part of IEEE 802.11 specifications in 1997. The purpose of this protocol was to provide the security to the wireless network equivalent to the security of a wired network at that time. The main features and characteristics of the WEP are as follows:[191]

- Use the scheme of shared secret keys consisting of two different lengths – 40 and 104 bits.
- Those keys – 40 and 104 bits – are combined with the initialization vector (IV) of 24 bits each and 64-bit key and 128-bit key, respectively.
- Other versions of bits supported by the WEP protocol include 152 and 256 bits.
- The WEP protocol uses two types of keys as listed here:
  - **Unicast session key** – It is used for the encryption of data between the wireless client and access point (AP).
  - **Multicast or global key** – This key is used for the protection of the multicast traffic sent out by the AP to the connected clients.
- WEP supports the following methods of authentications:
  - Shared key authentication
  - Open system authentication.
- All versions of the WEP protocol have been deprecated in 2004 and the predecessors of the WEP have taken over the entire wireless security managed by the WEP protocol.

WI-FI PROTECTED ACCESS (WPA)

As mentioned in the aforementioned topic, the WEP protocol had some serious flaws and vulnerabilities, which could be exploited to breach the security of the data over the wireless network. To overcome those flaws, a new standard with substantial modifications and enhancements was released by the IEEE in 2004 under the implementation of the IEEE 802.11i standard. This new security standard for wireless LAN networks was named WPA. The wireless devices running WEP standard could be updated through software/firmware upgrade easily.

The most fundamental changes from the previous wireless security standard (WEP) included the improvement in the encryption algorithm. WEP used RC4 encryption methodology with 64-bit and 128-bit encryption keys, but the WPA uses a 256-bit encryption key, which is double secure as compared to the previous one. A new feature to change the key in every packet was adopted through an automated protocol known as Temporal Key Integrity Protocol (TKIP). This protocol dynamically adds a 256-bit encryption key in every packet of data traveling over the wireless network.[192]

The Cyclic Redundancy Check (CRC) used in the WEP was replaced by the Message Integrity Check (MIC) to cover the flaws in the CRC method exploited for cyberattacks. The other changes and major features and characteristics of the WPA protocol are mentioned in the following:

- The TKIP protocol offers stronger security and integrity as well as the way for online firmware upgrade options.
- The WPA protocol supports two kinds of authentication modes as compared to the one in WEP protocol as listed here:
  - **Personal authentication mode** – Used for only personal authentication through a pre-shared key, which is not shared over wireless interface, but the client and APs use four-way handshake, which uses the pre-shared key for generating the encryption key.
  - **Enterprise authentication mode** – In this mode, a remote authentication server, normally a Remote Authentication Dial-in Service (RADIUS) is used.
- The encryption algorithm used by the WPA security protocol is still the same RC4 as used by the WEP protocol but at higher bit encryption.

This protocol faced certain security flaws and a new standard with advanced features known as WPA2 was introduced to make wireless security more robust and powerful.

WI-FI PROTECTED ACCESS 2 (WPA2)

WPA2 is the advanced version of the WPA wireless security protocol as explained in the aforementioned topic. This protocol was released in 2004 for replacing the existing WPA certifications on the wireless devices and the year 2006 was stipulated a cut-off date for mandatory WPA2 certified devices. The main advancements in the features, capabilities, functions, and characteristics of WPA2 are mentioned in the following list:[193]

- It uses the advanced encryption protocol named Counter Mode with Cipher Block Chaining Message Authentication Code Protocol. This protocol supports two methods:
  - AES counter mode encryption
  - Cipher Block Chaining Message Authentication Code.
- It uses AES cryptography instead of RC4 in the previous version.
- A new feature named Wi-Fi Protected Setup (WPS) was introduced.

A few vulnerabilities emerged till 2011 in the WPA2 such as "exposed pre-shared key" in the WPS feature, which led to the release of the advanced releases and version of this wireless security. Since 2018, the WPA3 has been released for even better wireless security.

EXTENSIBLE AUTHENTICATION PROTOCOL (EAP)

Extensible authentication protocol, precisely referred to as EAP, is a large family or framework of different kinds of authentication methods built over the EAP platform. This is the core framework used for the authentication of wireless users/clients to access computer and data resources over the wireless networks. As the name implies, it is an extensible protocol, which can be customized with a wide range of cryptographic and authentication mechanisms to develop a customized method for the authentication of genuine clients over wireless networks. This protocol is primarily used in wireless networks for authentication. This framework is adopted as the canonical authentication mechanism for wireless LAN networks defined under IEEE 802.11x standards.

The main features and characteristics of the extensible authentication protocol (EAP) are mentioned in the following list:[194]

- The current EAP framework is defined under the RFC 5247, which has replaced RFC 3748.
- Supports more than 40 methods for the authentication procedures, which are defined through different RFCs.
- Being an extensible protocol, a few vender-specific methods are also in use for different authentication applications.
- A few very common and popular methods include EAP-TLS, EAP-MD5, EAP-AKA, EAP-SIM, Cisco-LEAP, and many others.
- The EAP standard defines the message formats, and the supporting protocol defines the encapsulation of the messages in wireless communication.
- In the authentication methods powered by the EAP, the authenticator (server) initiates the authentication process rather than the client, which normally initiates the authentication process in majority of the other authentication protocols.
- The role of EAP in wireless authentication is to pass authentication information within the authenticator, which acts as the proxy of the authentication server, and supplicant, which is the software client on the wireless workstation.

EAP-TLS PROTOCOL

Extensible Authentication Protocol – Transport Layer Security, precisely known as EAP-TLS, is an authentication method or mechanism developed over the EAP framework. It is one of the most popular methods deployed for authentication purposes in wireless networks nowadays. This authentication protocol is standardized by the IETF through standard RFC 5216.

In this authentication method, the TLS protocol provides the following functions and activities:[195]

- Certificate-based mutual authentication of wireless elements
- The negotiation of key between the two endpoints in the wireless network
- The negotiation for the integrity-protected cipher suite

The authentication process in this mechanism starts with the negotiation for EAP between the peer and EAP authenticator. The authenticator sends an identity packet or

EAP request to the peer node, which responds with a reply known as EAP-Response or Identity Packet, which contains the user identity of the peer element. After this conversation, the authenticator will start just passing the messages between the peer and the authentication servers such as RADIUS, AAA authentication server, or Microsoft IAS server. The EAP server sends the EAP-TLS/Start packet, which has information such as start bit set (S) and the name of the type of authentication method, i.e., EAP-TLS. This packet does not have any other data. The communication between the authentication server and the peer starts from this packet and the peer starts responding to the messages.

The main features and characteristics of this method of EAP-TLS authentication protocol are mentioned here:

- It is an open standard defined by IETF supported by a large number of wireless devices vendors.
- It is one of the most robust and powerful security/authentication mechanisms available in the marketplace at present.
- It supports the client-side certification, which is the core element for such as robust security.
- It supports the storage of private keys on the smart cards, where the key is stored in encrypted code and authenticated by a PIN code.

PROTECTED EXTENSIBLE AUTHENTICATION PROTOCOL (PEAP)

The protected extensible authentication protocol or precisely known as PEAP protocol is an encapsulating protocol for different types of EAP-based methods, especially EAP-TLS protocol for authentication. It was designed to cover the deficiencies in the EAP protocol so that the additional security to the EAP messages is achieved through encapsulation and tunneling. The protocol was developed by the combined team of Microsoft, Cisco Systems, and RSA Security.

The main purposes, objectives, and characteristics of the Protected EAP protocol are mentioned in the following list:[196]

- Provides a secure method for transporting the authentication information based on the EAP-TLS protocol over the secure channel developed through encapsulation. The authentication may include usernames, passwords, digital certificates, private/public keys, and other information.
- Establishes a tunneling mechanism between the PEAP authentication servers and PEAP clients in the wireless network environment.
- Uses the server-side certificate for the authentication purpose and does not require a digital certificate on the client side.
- It supports WEP key management feature but does not detect any rogue AP.
- Supports the mutual authentication attributes and offers a high level of wireless security.

TRANSPORT LAYER SECURITY (TLS)

Transport Layer Security, precisely known as TSL, is a security protocol based on the encryption technology to provide security to the integrity and the privacy of data or communication packets traveling over the Internet. The security is achieved by encrypting the packets of communication transporting between two hosts or two applications.

The transmission of encrypted data and communication packets makes it very difficult for eavesdroppers and hackers to intercept, decode, and understand the information wrapped within the encrypted code. The TSL security protocol can be used in a wide range of applications such as web applications, file transfers, emails, audio and video data transporting in the real-time environment, text messages, instant messages, and other applications. It is also very important to note that the DNS and NTP servers also use this protocol for secure communication over the Internet.

The main features and characteristics of TLS protocol are mentioned in the following list:[197]

- The TLS protocol is based on the SSH protocol, which was adopted in 1994; later on, it was replaced by the TLS security in 1999 through an IETF adopted RFC 2246.
- The latest version of this protocol is known as TLS v1.3 released in 2018.
- The TLS protocol works above the TCP protocol and encrypts the data at the application layer such as HTTP, SMTP, IMAP, and others.
- TLS protocol can also be implemented on other transport-level protocols such as UDP, SCTP, DCCP, and others for real-time modern applications.
- The TLS protocol implements both symmetric and asymmetric cryptography to achieve a better tradeoff between performance and security.
- Supports a wide range of key-generating and exchange techniques such as ECDHE, DH, RSA, DHE, ECDH, and others.
- The main domains that the TLS protocol deals with in terms of security include the following:
  - **Data Encryption** – To save data hacked by the third party
  - **Data Integrity** – To ensure the sending and receiving of uncompromised data
  - **User Authentication** – To ensure the communicating parties are genuine and verified.
- The TLS connection starts through an authentication process known as TLS handshake.
- The implementation of TLS security on your web environment is done by using the valid SSL certificate issued by the authorized entities known as Certification Authorities (CAs).

ENCAPSULATING SECURITY PAYLOAD (ESP) PROTOCOL

Encapsulating Security Payload is a domain-specific security protocol, precisely known as ESP protocol. It is extensively used with the IPsec security protocol for making the payloads/data secure by encrypting the data (not the packet headers in the IPsec environment). It is an IETF-adopted security protocol for the payloads in the secure transmission of data over the Internet. It is released under the RFC 1827. It provides better confidentiality and integrity to the IP datagrams and in some cases, it is also capable of providing the authentication capabilities to the IP datagrams.

The main features and characteristics of the ESP protocol are mentioned here: [198]

- It supports both versions of IP, i.e., IPv4 and IPv6.
- It is scoped with the IPsec protocol for building a robust security.
- It is designed for the security of payload/data, not for the packet headers; but in certain cases, like tunneling environment, the entire packet can be encrypted as a payload under the jurisdiction of ESP protocol.
- The main components or fields of an encapsulating security protocol header are mentioned in the following list:
  - Sequence number field
  - Security parameter index (SPI)
  - Next header field
  - Payload data field
  - Integrity check value field
  - Next header field
  - Padding field
  - Sequenced number field.
- The ESP protocol also supports two major modes of operations as mentioned here:
  - ESP in Tunnel Mode
  - ESP in Transport Mode.
- It is a very important transport-layer protocol used in the IPsec-powered protocol stack along with the authentication header (AH).

LAYER 2 TUNNELING PROTOCOL (L2TP)

Layer 2 Tunneling Protocol, precisely known as L2TP protocol, is a protocol that develops a virtual tunnel between two communication hosts or network applications. This protocol is extensively used for the VPN services between the two remote stations. This protocol does not include the implementation of encryption by itself for the security and integrity of the data traveling through the logical tunnel created over the public network. It uses the IPsec protocol for the purpose of encryption to offer more robustness in terms of data integrity and confidentiality.

The most salient features and characteristics of L2TP are mentioned in the following list:[199]

- This protocol looks like a protocol working at the data link layer of the OSI model, but in fact, it is a session layer protocol, which creates a logical data link (Tunnel) between the two endpoints in a point-to-point fashion.
- It uses the UDP port for communication at the transport layer of the OSI model.
- It is also known as the virtual dial-up connection between the ISP and the users for a dedicated logical connection over the public network.
- This protocol is published as an extension of point-to-point tunneling protocol (PPTP) in 1999.
- The L2TP protocol takes the characteristics of both Cisco Layer 2 Forwarding (L2F) and Microsoft's PPTP.
- It offers a cost-efficient, secure, reliable, and high-performance service for connecting multiple locations simultaneously.

MOBILE APPLICATION PART SECURITY (MAPSEC)

Mobile Application Part Security, precisely known as MAPSec, is a security protocol implemented over the SS7 signaling for a wide range of mobile services offered by the MAP communication protocol. When MAP protocol is used with the SS7 signaling, the security of the messages transported within the services, nodes, and users would be handled by the MAPSec protocol. If the MAP protocol is implemented over the IP protocol, the security of the messages is catered to by the IPsec security protocol.

The main features of MAPSec Protocol are mentioned in the following list:[200]

- Implements the encryption over the communication/signaling messages generated by the Mobile Application Part (MAP) protocol in the cellular mobile services.
- MAPSec is implemented between the services nodes, which communicate over the MAP protocol in cellular networks.
- It is an end-to-end message security protocol between two nodes other than the intermediate nodes of the services in the network.
- It supports six different types of profiles with different features and capabilities related to the security of the signaling.

- This protocol provides the following security services:
- Protection against the anti-replay
- Integrity of data/messages
- Authentication of data origination
- Confidentiality (optional feature).
- The most important purpose of the MAPSec protocol is to provide security to the MAP signaling messages in the SS7 signaling environment.

END-TO-END MESSAGE SECURITY
PROTOCOL (ENDSEC)

As mentioned in the aforementioned topic, the MAPSec protocol offers security between two nodes and it does not add security between multiple nodes to provide end-to-end security between two nodes including the intermediate nodes. For achieving better security, in MAP-specific signaling messages, an additional encryption is included to provide a complete end-to-end security, and with this, the EndSec protocol has been introduced. This protocol offers the capability to secure the signaling messages based on the MAP protocol in cellular mobile systems in terms of integrity, authentication of origination, and confidentiality at a higher level. This mechanism adds cryptographic checks into the MAP-originated signaling messages. Those checks automatically find the corruption in the messages and can automatically correct those messages. These also notify the root causes of the problem(s) and recognize nodes that are causing that corruption/problem in the communication.[201]

Sample Questions and Answers for Chapter 9

Q1. What are the two types of keys used by WEP protocol?

A1. WEP (Wired Equivalent Privacy Protocol) uses two types of keys as listed here:
- **Unicast session key** – It is used for the encryption of data between the wireless client and access point (AP).
- **Multicast or global key** – This key is used for the protection of the multicast traffic sent out by the access point to the connected clients.

Q2. What is EAP?

A2. Extensible Authentication Protocol, precisely referred to as EAP, is a large family or framework of different kinds of authentication methods built over the EAP platform. This is the core framework used for the authentication of wireless users/clients to access computer and data resources over the wireless networks.

Q3. Name the two major modes supported by Encapsulating Security Payload (ESP) protocol.

A3. The ESP protocol supports two major modes of operations as mentioned in the following:
- ESP in Tunnel Mode
- ESP in Transport Mode

Q4. State at least three features of L2TP.

A4. Three most salient features of Layer 2 Tunneling Protocol (L2TP) are:
- This protocol looks like a protocol working at the data link layer of the OSI model, but in fact, it is a session layer protocol, which creates a logical data link (Tunnel) between the two endpoints in a point-to-point fashion.
- It uses the UDP port for communication at the transport layer of the OSI model.
- It is also known as the virtual dial-up connection between the ISP and the users for a dedicated logical connection over the public network.

Q5. What are the security services provided by Mobile Application Part Security (MAPSec) protocol?

A5. MAPSec protocol provides the following security services:
- Protection against the anti-replay
- Integrity of data/messages
- Authentication of data origination
- Confidentiality (optional feature)

Server Level Security Systems 10

With the growth in the network users, it became a difficult task for any node to perform multiple tasks simultaneously with the efficiency and performance required for the modern network services. To improve the performance of the networks, specialized entities for authentication, authorization, accounting, security, privacy, and other security purposes were introduced into the network so that the better performance of the network may be maintained. Numerous specialized network nodes for security and authentication purposes were developed and implemented in the network so that a large number of customers or end-users can easily be catered to without compromising on the performance and quality of the services.

Some of the very important security and authentication services performed by the dedicated servers in the network, especially in the Internet environment, are mentioned in this particular chapter for easy understanding.

REMOTE AUTHENTICATION DIAL-IN USER SERVICE (RADIUS)

Remote Authentication Dial-In User Service, precisely known as RADIUS, is a protocol to handle remote authentication in a large-sized network. It is a networking protocol based on the client and server mode of communication. It offers a comprehensive user management of network resource users for authentication, authorization, and accounting activities associated with the user profiles for certain network services.

The RADIUS server protocol is exclusively designed for the network services based on certain web applications in which the users access the network resources and the functionalities of IP mobility from anywhere in the world. It plays a very vital role in modern

DOI: 10.1201/9781003300908-13

telecom as well as web-based services offered by the large internet service providers and telecommunication service providers across the globe. This server-level network protocol uses different other protocols for encryption as well as for the other types of security functions.

For instance, when a user wants to use any service that it is subscribed to, it sends a request to the service handling server or telecom node; the service handling node is normally referred to as Network Access Server (NAS). The NAS forwards this request to the RADIUS server for authentication, authorization, and accounting verification of that particular user. If the user is a valid client registered with the service and has all other valid attributes to access the service, it sends an acceptance message to the NAS device, which allows the user to connect to the network services to use the services that the client wants to. The recording of the billing is also maintained by the RADIUS server protocol.

The main features of the RADIUS network protocol are mentioned here:[202]

- This protocol is used as a proxy server in the network service environment for AAA functionalities for a remote user for the other servers too.
- It is an extensible protocol so the service providers can customize and use this protocol as per their own needs.
- It is a stateless protocol and works over the UDP port numbers 1812 and 1813.
- For the remote monitoring purpose, it uses the SNMP.
- It uses authentication of the remote users; it uses the following protocols:
  - Password Authentication Protocol (PAP)
  - Challenge Handshake Authentication Protocol (CHAP)
  - Extensible Authentication Protocol (EAP)
  - PPP
  - Simple Unix-based logins.
- This protocol is capable of looking into Lightweight Directory Access Protocol (LDAP) for accessing database information file.
- This protocol communicates with the NAS server for granting or denying the access to the remote user in the network.
- The session monitoring is accomplished by the RADIUS protocol that is ultimately used for billing purposes.
- The NAS server is also referred to as the RADIUS client in the network terminology.
- The format of the RADIUS packet consists of the following fields:
  - **Code field** – It is a one-byte field to identify the type of packet.
  - **Length field** – It is a 2-byte field, which defines the length of the packet.
  - **Identifier field** – It is a 1-byte field to identify/differentiate the packets.
  - **Authenticator field** – This field is used for filling some requests and responses. It consists of 16 bytes of space.
  - **Attribute field** – This field defines different types of attributes supported by the Radius server. There are over 63 attributes supported by the RADIUS protocol.
- Passwords are always sent encrypted in the access-request message from the client in the RADIUS protocol-based communication.

INTERNET AUTHENTICATION SERVICE (IAS)

Internet Authentication Service, shortly known as IAS service, is an implementation of limited features and capabilities of RADIUS protocol in the Windows operating systems such as Windows NT, Windows Server 2000, Windows Server 2003, and the later versions. It is used for the authentication, authorization, and accounting management of remote access to a host running Windows operating systems. It is implemented as a service component in the OS platform.

The IAS service supports only the Windows-based clients as compared to the RADIUS protocol, which supports all other third-party clients in the network. In the IAS environment, the user information is added into the Active Directory, which is accessed by the IAS service to authenticate the remote user for granting access to the remote host. After the initial introduction of this service in Windows NT, advanced features like the resolution of user name, which are parts of Windows server domain, additional security, and support for UTF-8 logging, were introduced in the Windows 2000 version.

The later version of Windows 2003 supported the advanced login feature through MS SQL Server and cross-forest trust relationship. The port-based login authentication for the IEEE 802.1X standards was also introduced in this version. The main features and characteristics of Internet Authentication Service (IAS) are mentioned here:[203]

- IAS is included in the Windows operating systems as a proxy of the RADIUS protocol.
- It is a proprietary network service developed by Microsoft Corporation.
- IAS communicates with the client through Remote Access Servers (RAS), which act as a proxy for a client in the network to communicate.
- This service uses the RADIUS protocol for communication for authentication and authorization.
- This service applies to only the dial-in services in any telecom or local networks.
- The name of the IAS server has been changed now; it is renamed Network Policy Server (NPS).
- The IAS server supports the following APIs:
  - Server Data Objects API
  - Network Policy Server Extension API.

The renamed version of IAS service known as Network Policy Server has emerged in Windows 2008 (and the advanced versions). It supports all the basic functions and capabilities of the IAS with numerous additional services and capabilities that make it more secure, reliable, and robust in terms of network security. A few very important additional features of NPS service are mentioned in the following list:

- NPS acts as the central server for the protection of the network access known as Network Access Protection (NAP).
- It supports the policy-based profiling of any remote user to access the network services.

- NPS supports the EAPHost as an extensible component of EAP policy.
- It supports both IP version 6 (IPv6) and IP version 4 (IPv4).

Other than the aforementioned additional features, many others are also part of this new version of authentication, authorization, and accounting service integrated with the modern Windows operating systems.

VIRTUAL PRIVATE NETWORK (VPN)

Virtual Private Network, commonly referred to as VPN, is a technology that provides security, anonymity, and privacy over the public network with the help of numerous security and encryption protocols either in combination or in a stand-alone implementation. In simple words, VPN is a secure and private connection in a public network in which the nature, origin, and other attributes of data as well as communication are hidden and encrypted from the hackers and the service providers in a public networking service domain.

It is an encrypted connection powered by one or more than one different security and encryption protocols over the public network like the Internet. Both the data and communication over the VPN network are free from eavesdropping and breach of privacy. This technology is vastly used in the corporate environment for remote work-from-home applications. The VPN services can be categorized into two major classes as mentioned here:[204]

- Remote Access VPN
- Site to Site VPN

In the first type, the remote device away from the headquarters or head office can be connected securely and privately over the Internet for accessing the corporate network. On the other hand, in the second type of VPN network, two remote branch offices located in different cities can be connected securely and privately over the public network like the Internet or other core network(s). In both cases, the connections are highly secure and free from any kinds of breaches associated with the public networks.

The main features and characteristics of VPN technology are mentioned here: [205]

- The entire traffic and data are fully encrypted and secure while traveling over the public network within the VPN connections.
- It supports the device posture features to verify if the device that is being connected through the VNP network is qualified enough to remain secure and reliable in the environment.
- It uses different types of security and encryption protocols either in a single protocol configuration or in multiple protocol configurations. The major types of security/encryption protocols used by the VPN technology are listed here:
  - **PPT Protocol** – Point to Point Tunneling Protocol (PPT) is a proprietary VPN protocol developed by Microsoft in the early 90s. It was introduced in Windows 95 for dial-up users to log in over a fully encrypted connection.

- **L2TP/IPsec Protocol** – Layer 2 Tunneling Protocol (L2TP) and IPsec protocol are used in combination for materializing a robust and faster VPN connection. The tunneling protocol handles the way traffic travels securely and IPsec handles the encryption of the data and communication over the network connection.
- **SSTP Protocol** – This is another very important protocol used in VPN technology. It is also integrated into the Microsoft operating systems. It uses a 256-bit SSL key for encryption and 2,048-bit SSL/TLS digital certificate for authentication.
- **OpenVPN Protocol** – This is an open-source protocol. It uses 2,048-bit RSA authentication, 256-bit AES encryption, and 160-bit SHA1 hash protocols.
- **IKEv2 Protocol** – Internet Key Exchange v2 protocol is used for key exchange session over the encrypted connection. It is paired with other security protocols like IPsec and others in the configuration of VPN connections.

The popularity of VPN networks has increased manifold in recent years due to the increasing threats to data and communication in the modern era of the Internet, especially in the domain of corporate and online businesses.

REMOTE DESKTOP SERVICE (RDS)

Remote Desktop Service, precisely RDS, is an integrated service in the Windows operating system 2008 and earlier. It is used for accessing remote computers virtually through a client. This service is based on the Remote Desktop Protocol, precisely known as RDP. This protocol handles the authentication, authorization, and accounting of a session and monitors the security and other parameters of the communication.

The main features and characteristics of RDP protocol and RDS service are mentioned here:[206]

- Microsoft Corporation's proprietary service integrated into the Windows Server operating systems 2008 and earlier.
- It is a thin client architecture implementation of the Microsoft operating systems.
- The accessed machine through RDS acts as a server and all commands are executed on that machine from the client software such as RemoteFX remotely.
- It supports the following major client components:
  - Remote Desktop Connection (RDC)
  - Windows Remote Assistance (WRA)
  - Win Logon's Fast User Switching (FUS)
- The most common server component of the RDS service powered by the RDP protocol is known as a terminal server (TS).
- Terminal server connects with the client over the TCP port 3389.

- This service supports a wide range of roles such as remote desktop gateway, remote desktop session host, remote desktop web access, remote desktop licensing, and others.

DOMAIN NAME SYSTEM (DNS)

Domain Name System, commonly referred to as DNS system, is an architecture of distributed network servers that provide the IP addresses against the names of the websites in the networked world or the Internet across the globe. It is a secure system for resolving the names into the IP addresses of the servers, which are defined through a unique IP address, which is the core component in the transmission of the data in the TCP/IP protocol suite.

When we try to access a website of any organization or entity on the Internet, we type in the uniform resource locator or URL of that website, which is like *organization.name. com*. A common user can understand and remember the URL easily rather than remembering the similar-looking IP addresses of a wide range of websites. But, the URL is a text-type presentation of the website address, which is not understood by the computers on the Internet. The computers use the IP addresses for locating and communicating with the clients on the Internet. To translate the URL address into an IP address, the role of DNS comes into play. The client sends a request to the DNS server for translating it into the IP address so that the communication can take place. The DNS server looks up the IP address against the URL address and replies with the IP address to the client for further routing of the communication. So, we can say that DNS is like an address book or telephone directory we use in our day-to-day work manually. Maintaining the addresses of millions of websites and IP addresses manually is not a practical stuff. So, an automated system was developed for finding the IP address against the website URLs that is known as the DNS system. The schematic diagram of how does a DNS system works in the internet environment is shown in Figure 10.1.

The main features and characteristics of DNS are mentioned here:[207]

- DNS is an application-specific protocol for IP address resolution in the networked world.
- It was developed by Paul Mockapetris in 1983.
- It was adopted by the IETF through RFC 882 and RFC 883 in November 1983, shortly after its development.
- It is a distributed directory service of the global networked world based on the Internet.
- DNS system is based on two types of servers as listed here:
  - **Recursive DNS server** – Also known as a recursive resolver, which is operated by the local ISPs or local authorities. It is the first server to cater to your request.
  - **Authoritative DNS server** – This is the real DNS server that has the entire information about the URLs and IP addresses run by the concerned Internet authorities.
- This system consists of a few major database records such as A record, TXT record, Canonical Name (CNAME) record, Mail Exchange (MX) record, and others.

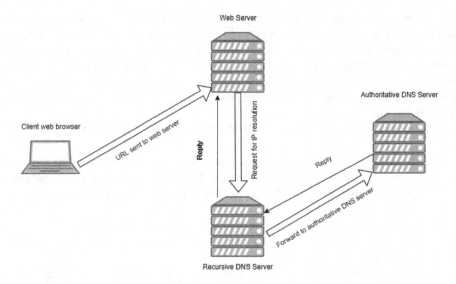

FIGURE 10.1   Schematic diagram of DNS architecture.

- It uses the TCP/IP suite underneath and uses the UDP on port number 53 at the transport layer of the OSI model of communication.
- It uses two types of messages that have the same format. Those messages include:
  - **Query message** – Initiated by the client
  - **Reply message** – Initiated by the server in response to the query message.
- Each message used in the DNS communication has four sections named:
  - Question section
  - Answer section
  - Authority section
  - Additional space section
- Each message has one header other than the aforementioned four sections. There are eight different types of fields in the DNS message header.

The role of DNS system is very vital in the entire communication over the internet. It offers a solution to a very complex problem in the gigantic network of the networks along with the systematic security features and capabilities.

ADDRESS RESOLUTION PROTOCOL (ARP)

Address Resolution Protocol, concisely referred to as ARP protocol, is the lower layer of data link protocol. This protocol is used for translating the IP address associated with the MAC address, which is also known as the physical address or hardware address of the computer interface or host. This protocol is exclusively used for IPv4 addresses. For the IPv6 addresses, another data link layer protocol named NDP is used.

The mapping of the Internet address to the MAC address is a very critical part of the entire TCP/IP protocol suite as well as in many other protocols of layer 2 in the OSI model of communication. This protocol is used for the transmission of the frames within a network. The ARP protocol was adopted in the year 1982 through RFC826, which is an Internet Standard named STD 37. This protocol can be implemented in a wide range of protocols such as the entire family of IEEE 802 standards, FDDI protocol, DECnet, ATM, Frame Relay, and others.

The main features and characteristics of ARP are mentioned in the following list:[208]

- The ARP protocol is based on the request–response mode of communication.
- The working jurisdiction of ARP protocol is within the local network and the scope of routing of the ARP messages is not valid across the network or router of the networks.
- It works within the boundaries of the data link layer of the OSI model.
- The network layer protocol (IPv4) and ARP work very closely at Layers 2 and 3 of the network.
- Any host which knows about the MAC address to whom it has communicated in recent times, the MAC address of that host is stored for future use.
- If any host does not have the MAC address of the destination node, it will broadcast a frame asking for the MAC address of the machine whose IP address is known. This IP address is sent in the broadcast message known as ARP Request. All machines will check the message and discard if they are not the machine with the said IP address. The only machine whose IP address is broadcast in the message replies the query with an ARP-Response. This response will be unicast to the machine that has sent the broadcast messages. All other machines will discard the broadcast message after reading its irrelevancy with them.
- Once the source machine knows the MAC address of the machine that has replied with the ARQ-Response would be saved in the ARP Cache for future use for a certain period, which is known as ARP Cache Timeout period.
- Another protocol known as Reverse ARP is used for determining the IP address of the host whose MAC address is known. This protocol also works in the same way with a little difference in the same layers of communication.
- The ARP Cache is supported by all operating systems used in modern communication based on packet transmission (IPv4).
- The ARP protocol supports three basic techniques for address resolution as mentioned here:
    - **Table Lookup Technique** – A table is manipulated by the machine with the MAC addresses of all machines that have communicated recently. For any new transmission, first of all, that table is checked for the MAC address.
    - **Dynamic Technique** – This is the case when the MAC address is not available in the table. An ARP request is broadcast to get the response dynamically at the time of need.
    - **Closed-Form Computation** – This method a hardware protocol address based on the hardware address is used.

- There are four types of ARP protocols commonly adopted in the communication as mentioned in the following list:
  - Reverse Address Resolution Protocol
  - Proxy Address Resolution Protocol
  - Inverse Address Resolution Protocol
  - Gratuitous Address Resolution Protocol.
- The header of an ARP message – either request or response – consists of the nine fields as mentioned here:[209]
  - **Hardware Type (HTYPE)** – This field specifies the type of network link protocol. For instance, the value for Ethernet protocol is 1.
  - **Protocol Type (PTYPE)** – This field specifies the internetwork protocol such as IPv4, IPX, DDN, or any other protocol for which the ARP-Request is meant.
  - **Hardware Length (HLEN)** – This field specifies the length of the hardware address.
  - **Protocol Length (PLEN)** – The protocol length field specifies the value in bytes for the protocol type such as IPv4 and others.
  - **Sender Hardware Address (SHA)** – The definition of the address of the hardware of the host that has initiated the ARP message.
  - **Sender Protocol Address (SPA)** – This field defines the address of the device used in the internetwork protocol.
  - **Operation (OPER)** – The specification of operation types such as ARP-Request or ARP-Response is done through this field.
  - **Target Hardware Address (THA)** – This is the MAC address of the receiving hardware. This field applies to the ARP-Response only. The requests are broadcasts; so, no destination MAC is not possible to add.
  - **Target Protocol Address (TPA)** – The protocol address associated with the internetwork protocol is defined by this field.
- The values of protocol parameters are already defined and assigned by the IANA, which also deals with the assignment of IP addresses to customers across the globe.

NETWORK TIME PROTOCOL (NTP)

Network Time Protocol, commonly referred to as NTP protocol in the domain of IT and telecom, is a protocol for synchronizing the elements of a large network of diverse platforms, devices, and other elements and applications through a uniform time source. This is one of the oldest protocols in the field of communication that is still working effectively in all types of networks powered by different types of protocols. The use of NTP in the existing Internet based on TCP/IP protocol suite is very widespread and very important in running the entire system in a synchronized manner.

This protocol was initially designed by David L Mills, a researcher at the University of Delaware, USA. This protocol was put into operations well before 1985 when it was publicly adopted. Numerous versions of this protocol have been developed and adopted by the IETF. This protocol is officially adopted by the IETF under the RFC 1305.[210]

Initially, the synchronization protocol was put into the operations in the name of the DCNET Internet clock service in 1980. In 1985, it was officially regularized under the RFC 958, and then, it has been updated and modernized with additional features and capabilities through multiple RFC documents such as RFC 1059, RFC1305, FRC5905, and others.

The main features and characteristics of NTP are mentioned in the following list:[211]

- The reference time used in the NTP protocol is Coordinated Universal Time (UTC).
- For the right selection of the time server, this protocol uses different types of algorithms; a few of them are mentioned here:
  - Marzullo's algorithm
  - Intersection algorithm.
- The time tolerance ranges within the 10s of the milliseconds on the Internet of the networks.
- This protocol is based on the client–server model of communication. But in certain cases, it can be implemented in peer-to-peer configuration too.
- It uses the UDP protocol and the UDP port number 123 in the network based on the TCP/IP protocol suite.
- The NTP protocol uses the hierarchical sources of time sources structured in semi-layered architecture for drawing the time.
- It uses four types of strata of time, as mentioned in the following:
  - **Stratum 0** – This layer or hierarchy devices are those which are highly accurate in producing time source and are commonly used as the NTP time source/servers.
  - **Stratum 1** – They are normally directly connected to stratum 0 devices and are known for being the second-highest accurate time sources and used as stratum 1 servers. They are synchronized within a few microseconds of their stratum 0 devices.
  - **Stratum 2** – Network devices synchronized with the Stratum 1 servers are also known as stratum 2 servers.
  - **Stratum 3** – They get synchronized with the stratum 2 servers and peer with the same layered machines and offer synchronization service to the stratum 4 devices, and so on.
- The NTP servers use UDP port number 123 and the NTP clients use the UDP port number 1023.

The NTP is also capable of identifying the time source based on the reference identities commonly known as REFIDs or Reference Identifiers. A huge list of REFIDs is adopted for that purpose.

Sample Questions and Answers for Chapter 10

Q1. What is RADIUS protocol?

A1. Remote Authentication Dial-In User Service, precisely known as RADIUS, is a protocol to handle remote authentication in a large-sized network. It is a networking protocol based on the client and server mode of communication. It offers a comprehensive user management of network resource users for authentication, authorization, and accounting activities associated with the user profiles for certain network services.

Q2. What are the APIs supported by the IAS server?

A2. The IAS server supports the following APIs (Application Programming Interfaces):
- Server Data Objects API
- Network Policy Server Extension API

Q3. What are the major client components that the Remote Desktop Service (RDS) supports?

A3. RDS supports the following major client components:
- Remote Desktop Connection (RDC)
- Windows Remote Assistance (WRA)
- Win Logon's Fast User Switching (FUS)

Q4. What are the two types of servers that a DNS system is based on?

A4. DNS system is based on two types of servers:
- **Recursive DNS server** – Also known as a recursive resolver, which is operated by the local ISPs or local authorities. It is the first server to cater to your request.
- **Authoritative DNS server** – This is the real DNS server that has the entire information about the URLs and IP addresses run by the concerned Internet authorities.

Q5. What does IANA stand for?

A5. Internet Assigned Numbers Authority (IANA)

Privacy

11

Privacy is a big concern in modern computing. We see houses or premises marked "No Trespassing", meaning only authorized personnel are allowed access to such locations. Anyone without authorized access entering those areas may be in violation of the law. Authorized Computer user also maintains private space and any unauthorized user with unauthorized access would be considered trespassing and in violation of agreed-upon rules and policies. Password is one of the largest privacy protocols.

As we entered Web 2.0 in this millennium, WWW, Internet, and social media have become very crucial to conduct day-to-day businesses. We also have noticed that during COVID-19 (Coronavirus disease, SARS-CoV-2) pandemic, almost the entire world conducted businesses using the Internet. In such a scenario, it is very important that every device must be secured with authorized access and privacy is to be maintained.

Privacy is a very broad topic. We would focus on some general guidelines of privacy in this chapter. We introduce the chapter by discussing the differences between privacy and security. The rest of the chapter is focused on discussing different types of privacy which may overlap as there is no strict guideline that has been set to define different types of privacy.

DOI: 10.1201/9781003300908-14

DIFFERENCE BETWEEN PRIVACY AND SECURITY

Privacy and security are interrelated. Privacy is more like information or data contained with personal information. For example, in the USA, social security number (SSN) is a very private personal number that should be only used for financial correspondence with SSN bearer's full consent. Similarly, credit card information, date of birth (DOB), residence, occupation, etc., are also personal information of an individual. Someone may share residence, occupation, etc., information to the public on social media platforms or websites where others might keep it private. Any personal information or data shared with the public is not private anymore (i.e., privacy is not maintained).

Anytime personal/confidential information or data shared with an organization should be protected from intruders through enabling the protective tools or mechanisms, which is the essence of security. It is the responsibility of the organization to secure private data. However, often organizations, especially social media platforms like Facebook, Twitter, Google, etc., disclose some private data such as someone's purchase behavior, Internet browsing history, likes versus dislikes, etc., to third-party vendors to make financial gains. This sharing is legal since the user agrees to the terms and conditions provided by such organizations. In such cases, privacy is not maintained anymore, but the security is still maintained by not sharing the sensitive information such as DOB.

In a nutshell, we can understand that user privacy is preserved utilizing security tools.

TYPES OF PRIVACY

It is rather difficult to pinpoint the types of privacy as privacy might differ based on the tools and usages. For example, Internet privacy and home privacy might not follow the same type. Also, privacy differs from person to person and one country to the other as laws and regulations differ for each country.

We will focus on the general three types of digital privacy mentioned in Wikipedia:[212]

- Information privacy
- Communication privacy
- Individual privacy

As mentioned earlier, these types or other subtypes are loosely defined as they may overlap into multiple types (there is no universal consensus on these issues). Hence, we will mention various types without maintaining any specific order.

INFORMATION PRIVACY

Information privacy is concerned with how any individual's data or information is collected or preserved. The idea is derived from Information Technology (IT) and the idea of information privacy emerged in the 1990s.[212]

Today, we are consuming so much information or data constantly. Data growth has become exponential in recent years. Everyday people digitally connected to the cloud or internal storage are interacting through the Internet or IoT or other devices that are producing a gigantic amount of data. It is expected that there will be 75 billion IoT devices worldwide by 2025[213], and at the same token, there will be 175 zettabytes of data in the global data sphere. One zettabyte is equal to one sextillion bytes or 1,021 (1,000,000,000,000,000,000,000) bytes, or one zettabyte is equal to a trillion gigabytes.

The privacy of how information is collected, preserved, and used in any organization is crucial. Both European Union (EU) and the USA have adopted multiple privacy laws in this regard. EU adopted relatively stricter guidelines to protect user rights on how a company may use user information whereas US laws are more relaxed and allow an individual organization to self-regulate the privacy of the user information.

DEFENSIVE PRIVACY

This privacy is to protect a person's sensitive information from unwanted individuals or organizations. Every individual usually wants to protect certain information to avoid vulnerability or risk from the outside world. Some examples of information or data that fall under defensive privacy are US tax returns, medical diagnosis, financial records, and so on. Such information may vary from location to location as well. Someone might want to protect his female colleague's or relative's identity from the outside world as it might not be safe to disclose their information to the public.

We need to maintain defensive privacy to safeguard us to be vigilant and protect us from the evil person who wants to cause harm to the victim and society in general. This is very important to be cautious when we share or post content on the Internet. The defensive privacy is not always maintained as US Supreme Court ruled in 1974[214] – individual information collected by the Bank belongs to that Bank and Bank is allowed to do anything as they deem necessary with that information.

HUMAN-RIGHTS PRIVACY

Human rights privacy refers to protecting information from a system or a governing body of the country. We know that lots of countries in the world including US spy on their private citizens. As we know since September 11 attack, the USA became very suspicious of Muslims (generally speaking). The government employed informants to spy on "Muslim Community" including spying inside religious institutions such as Masjid. Also, US President Donald Trump infamously banned 7 Muslim countries' citizens from entering the USA.

Human rights privacy is about government in violation of the privacy whereas defensive privacy is about an individual abusing the privacy.

PERSONAL PRIVACY

Personal privacy is protecting an individual's right which we discuss later in this chapter with its subtypes. Such privacy is protected under the US laws in the first, fourth, and fifth amendments in the US constitution where privacy is not defined clearly. However, the California Constitution guarantees its citizens, personal privacy.

This is to ensure the privacy of any individual in his/her personal space that is a person should not be disturbed or interrupted at his/her home without any valid reason.

CONTEXTUAL PRIVACY

This privacy is rather objectionable based on the location or person or organization. Sometimes sharing such information might not fall into privacy, whereas at other times, it might be harmful. For example, same-sex marriage is becoming more and more popular these days in the USA. It was prohibited even in the US military until recently. As same-sex marriage or relationships is becoming more accepted in the US society, it may not fall into privacy. However, in many other countries or locations around the world, it is illegal to have such a relationship where maintaining privacy on this is necessary.

COMMUNICATION PRIVACY

In the age of the Internet, there is constant communication between multiple parties using digital communication methods. Preserving the privacy of this communication between the sender and the receiver falls into this criterion. There are multiple ways that the communication can be intercepted or hacked.

Individuals or organizations constantly communicate through emails, video conferencing or phone, and other numerous media. Often these communications are intercepted. One recent example could be Zoom bombing that started during the COVID-19 pandemic time in 2020 where intruders interrupted numerous video conferencing sessions interrupting important corporate meetings or online classes for schools or universities. It is reported[215] that there were at least 50 zoom bombing per month that occurred during March 2020 and four months afterward. The popular communication interception is known as the man-in-the-middle (MITM) attack. According to Norton, MITM occurs by an intruder's interception in a communication between the sender and the receiver which can occur via multiple ways such as IP spoofing, DNS spoofing, HTTPS spoofing, SSL hijacking, Email hijacking, Wi-Fi eavesdropping, Stealing browser cookies[216]. The protection mechanisms are also mentioned on the Norton website like verifying HTTPS usage, avoiding of use of public Wi-Fi, staying away from phishing emails, and using a secure Wi-Fi connection.

Communication Privacy Management (CPM)[217] is introduced by Sandra Petronio in 1991 which explains why and how people would manage private disclosure. The CPM is important for privacy protection, preservation, and how the communication should happen. The subcriteria below are divided based on the CPM. CPM theory has been applied to social media and other platforms to identify intimate interpersonal relationships and interpersonal peer relationships.

PRIVATE INFORMATION

Private information differs on every individual's personality. Certain information may be private to one individual where others might not feel the same way. We also notice that a spouse may feel comfortable sharing certain information with the intimate partner. Often private information is shared between two spouses that could be of very sensitive nature as well.

Private information is more like a self-disclosure. It may also focus on someone sharing a secret with another person and not wanting the information to be revealed to anyone else.

PRIVATE BOUNDARIES

This is like drawing a line between what is considered private vs what is public. A person might not want to share certain information in an open forum discussion. We also see the US government often mentioned something to be classified meaning that falls into the private boundaries, to be shared only with the individuals who are authorized to have access. As we know, the classified information also has multiple levels and a person may have access based on the US security clearance and must be trusted and authorized by the US government.

CONTROL AND OWNERSHIP

The control and ownership fall into who gets to claim certain information of his/her own which is part of private and personal boundaries. If anyone's control or ownership is taken away, the person may feel threatened and insecure since the privacy of the information may not be preserved (like how the person wants it to be).

We may bring a US patent as an example of ownership where the individual or organization has the ownership of a certain invention. In such cases, this invention is shared with the authorization of the organization or the individual.

RULE-BASED MANAGEMENT

Rule-based management comprises privacy rule characteristics, boundary coordination, and boundary turbulence. It is a framework on how the decision is made concerning privacy. Rule-based management can be complex and can be adopted on an individual or entity level.

PRIVACY RULES/MANAGEMENT DIALECTICS

Privacy management is the opposing argument and the entity to reveal private information. The idea is how privacy is viewed by others and the society in general. Privacy rules are dictated by social norms and cultural bindings.

INDIVIDUAL PRIVACY

The privacy of a digital footprint on any individual refers to individual privacy. Digital footprint may refer to religious and political views, race, sexual orientation, personality, or intelligence which may be collected from individual's blog posts, social media interactions, and browsing logs. Social media intrusion is very popular these days which may cause violating individual privacy. Many employers in the USA have shaped the hiring criteria by visiting prospective employee's digital footprint. 75% US recruiters[218] conduct online search/research on the candidates through search engines, Twitter, social-networking sites, personal websites, photo/video-sharing sites, and blogs. On the other hand, it is also reported that human trafficking occurs in certain countries using YouTube and other social media platforms where minor children are often the victims.

Individual privacy may contain personally identifiable information (PII), which is any information related to identifying a person. There may be other subtypes concerning individual privacy. Without making a hierarchy for these, we will talk about more types/subtypes in the subsequent parts of this chapter.

PRIVACY OF THE PERSON (BODY, MIND, AND IDENTITY)

There are discussions on how any individual's privacy on different parts such as mind, body, and identity is preserved. Different countries in the world passed laws and regulations to maintain such privacy. This privacy comprises the privacy of any individual including the physical person, thoughts, autonomy, and identity. The following is the discussion on each of these subtypes.[219]

The protection or privacy of body is very important in every society. Every individual maintains a private space. No one is allowed to touch any individual without his/her consent. The consent may be verbal or signs or postures. One example of posture could be shaking hands of a stranger.

Privacy of mind or thought comprises freedom of conscience, thought and religion, and the freedom of expression. Majority of the countries in the world protect the privacy of mind under this context.

A personal decision-making or autonomy is widely used in the USA. This is to maintain *his/her life* based on individual choice. For Example, every citizen in the USA has the right to choose his/her spouse to marry. A person has autonomy in decision-making based on the body and other surroundings.

The identity is about how one individual portrays himself/herself to others or how he/she wants to be recognized by others. This privacy may come from protecting an individual's social status, honor, and reputation or how the person wants to maintain the social life. Certain people invest lots of money and time to be recognized and stay influential in the society.

PRIVACY OF PLACES AND PROPERTY AND COMPUTERS

Almost every country in the world protects the privacy rights of everyone's personal spaces and property. Some constitutions call it dwelling while others call home or house. Some countries such as Poland consider vehicles as part of a home which are constitutionally protected against unlawful entry or search. Germany and Italy consider that business premises fall under home and are linked to a person's private life. US fourth amendment secures the right of people and their belongings whereas UK and Canada also have similar guidelines in their constitutions. Therefore, we see that private life places are protected from intrusion while the definition of this place may vary based on the country.

Privacy protection of computing or *Informatic* privacy is a newly added constitutional right for certain countries since majority of the world population has moved to a digital era. German constitutional court gives the fundamental right to the confidentiality and integrity of computer systems since 2008. Italian constitutional court is about to add computers as part of a home which is known as "informatic home" or "informatic privacy". Canadian Supreme Court recognizes computing, cell phones, and other personal devices as protected under privacy since 2014 as lots of information is contained in such devices. The USA has only identified cell phones to be protected under the fourth amendment since 2014. US appellate courts raised privacy concerns on personal computers and put those under the fourth amendment. Computers or other devices are now considered a new place where private information is stored which falls under proprietary privacy, and communicational privacy since these devices are used to store, send, and receive data.

PRIVACY OF RELATIONS

The privacy of family life, social relations, communications, documents, and person (body, mind, and identity) falls into this criterion. We already discussed Person's (Body, Mind, and Identity) privacy and communication privacy. In this section, we will highlight some other types.

There is a distinction between family life and private life. Some individuals may want to protect some data from spouses or family members. Family privacy is about the freedom of sharing information with other family members and maintaining family ties. Family privacy gives the freedom to build and raise a family freely.

Private life is protected to build social relationship with other human beings. If there is no unlawful interference of communication, a person is free to enable, maintain, and deepen the relations with other people in a country or around the world. Privacy of social needs is categorized into self-actualization, status (or self-esteem), love or belonging, and safety physiological or biological needs.[220]

There is always privacy of documents or information. We know that each country maintains certain classified documents to maintain national security. We know that Edward Snowden is wanted in the US court for disclosing the US classified documents to the public. Also, WikiLeaks is under threat for releasing classified documents from different nations. There is a big debate between freedom of speech and classified documents. Some governments violate the freedom of speech and classify certain documents to serve their own interests (corporations or certain individuals).

PRIVACY OF PERSONAL DATA

Data privacy as well as privacy of personal data is very popular as technology is moving forward and there are intruders and hackers who are trying to access these data through the network. In US or western countries, the medical data of a person are very much protected and not available to any individual except with the consent of the patient. Every business and institution guarantees the protection of its user data. The universities also do not share any grade with any individual without the student's consent. The PII such as SSN or DOB is very sensitive in the USA and cannot be shared with any individual without the bearer's consent.

We know there are so many personal data breaches recently, and some are reported here:[221]

- 533 million Users – Facebook, 03 April 2021
- Over 2 billion – BlueKai, 19 June 2020
- 5 billion – Keepnet Labs, 9 June 2020
- 8.3 billion – AIS, 25 May 2020
- 5.2 million – Marriott, 31 March 2020
- 250 million – Microsoft, 22 January 2020
- 267 million – Facebook, 19 December 2019
- 1 million – T-Mobile, 22 November 2019

There are numerous discussions still continuing on privacy issues as social media and Internet use are at the peak. People are sharing their ideas and thoughts via multiple platforms nowadays. We also have seen that certain governments and institutions are/ were abusing and even prosecuting the citizens to implement certain agendas or establish the government's ideology. As time progresses, we may see some better-defined universal guidelines on privacy issues.

Sample Questions and Answers for Chapter 11

Q1. What is information privacy?

A1. Information privacy is concerned with how any individual's data or information is collected or preserved. The idea is derived from Information Technology (IT) and the idea of information privacy emerged in the 1990s.

Q2. What is CPM?

A2. Communication Privacy Management (CPM) is introduced by Sandra Petronio in 1991 which explains why and how people would manage private disclosure.

Q3. Define Rule-Based Management of privacy.

A3. Rule-Based Management comprises privacy rule characteristics, boundary coordination, and boundary turbulence. It is a framework on how the decision is made concerning privacy. Rule-based management can be complex and can be adopted on an individual or entity level.

Q4. What is privacy of mind or thought?

A4. Privacy of mind or thought comprises freedom of conscience, thought and religion, and the freedom of expression. Majority of the countries in the world protect the privacy of mind under this context.

Q5. Give an example of sensitive information as may be considered in the USA.

A5. The personally identifiable information (PII) such as social security number (SSN) or date of birth (DOB) is very sensitive in the USA and cannot be shared with any individual without the bearer's consent.

References

1 Amer Al-Canaan, "Telecommunications Protocols Fundamentals," Chapter in *Telecommunication Systems – Principles and Applications of Wireless-Optical Technologies*, ISBN: 978-1-78984-293-7, 2019, DOI: 10.5772/intechopen.86338.
2 https://study.com/academy/lesson/the-components-of-a-telecommunications-system.html
3 Kutub Thakur and Al-Sakib Khan Pathan, *Securing Mobile Devices and Technology*. ISBN 9781032136127, CRC Press, Taylor & Francis, Boca Raton, FL, USA, 2021.
4 www.history.com/news/who-invented-the-internet
5 https://en.wikipedia.org/wiki/Claude_Chappe
6 https://en.wikipedia.org/wiki/Drums_in_communication
7 https://en.wikipedia.org/wiki/Smoke_signal
8 http://kotsanas.com/gb/exh.php?exhibit=1201001
9 http://scienceline.ucsb.edu/getkey.php?key=4026
10 www.history.com/topics/inventions/telegraph
11 https://en.wikipedia.org/wiki/Radio_frequency
12 https://searchnetworking.techtarget.com/definition/OSI
13 www.webopedia.com/quick_ref/OSI_Layers.asp
14 www.studytonight.com/computer-networks/osi-model-presentation-layer
15 www.studytonight.com/computer-networks/osi-model-network-layer
16 https://en.wikipedia.org/wiki/Data_link_layer
17 https://study.com/academy/lesson/physical-layer-of-the-osi-model-definition-components-media.html
18 www.techopedia.com/definition/8866/physical-layer
19 https://en.wikipedia.org/wiki/Modulation
20 https://ewh.ieee.org/reg/7/millennium/radio/radio_differences.html
21 www.spincore.com/products/PulseBlasterDDS-300/Modulation_Techniques/analog_modulation.shtml
22 "Introduction to Digital Modulation," Lecture slides of Murat Torlak, available at: https://personal.utdallas.edu/~torlak/courses/ee4367/lectures/lecturedm.pdf
23 https://en.wikipedia.org/wiki/Amplitude_modulation
24 www.sciencedirect.com/topics/physics-and-astronomy/frequency-modulation
25 www.pcmag.com/encyclopedia/term/phase-modulation
26 www.usna.edu/ECE/ec312/Lessons/wireless/EC312_Lesson_23_Digital_Modulation_Course_Notes.pdf
27 www.rohm.com/electronics-basics/wireless/modulation-methods
28 www.sciencedirect.com/topics/engineering/pulse-amplitude-modulation
29 www.elprocus.com/pulse-code-modulation-and-demodulation/
30 www.sciencedirect.com/topics/engineering/pulse-width-modulation
31 www.tutorialspoint.com/digital_communication/digital_communication_frequency_shift_keying.htm
32 https://en.wikipedia.org/wiki/Phase-shift_keying
33 www.sciencedirect.com/topics/engineering/quadrature-amplitude-modulation
34 https://teachcomputerscience.com/simplex-half-duplex-full-duplex/
35 www.tutorialspoint.com/analog_communication/analog_communication_multiplexing.htm
36 www.tutorialspoint.com/frequency-division-multiplexing

37 www.polytechnichub.com/advantages-disadvantages-applications-time-division-multiplex ing-tdm/
38 https://findanyanswer.com/what-is-multiplexing-and-explain-different-types-of-multiplexing?
39 https://en.wikipedia.org/wiki/Telegraph_code
40 https://pdfs.semanticscholar.org/10c5/1a7fa123e1911b60021dbda359a9e61e6a84.pdf
41 https://en.wikipedia.org/wiki/Wigwag_(flag_signals)
42 www.wikiwand.com/en/Optical_telegraph
43 www.britannica.com/biography/Samuel-F-B-Morse
44 www.ericsson.com/en/about-us/history/changing-the-world/phones-for-everyone/ the-invention-of-the-telephone
45 https://spectrum.ieee.org/telecom/standards/morse-codes-vanquished-competitor-the-dial-telegraph
46 https://en.wikipedia.org/wiki/Cooke_and_Wheatstone_telegraph
47 www.britannica.com/topic/Baudot-Code
48 https://billtuttememorial.org.uk/codebreaking/teleprinter-code/
49 www.computerhope.com/jargon/a/ascii.htm
50 www.ascii-code.com/
51 https://en.wikipedia.org/wiki/History_of_the_telephone
52 www.britannica.com/technology/telephone/Signaling#ref1117784
53 https://en.wikipedia.org/wiki/Pulse_dialing
54 https://en.wikipedia.org/wiki/Dual-tone_multi-frequency_signaling
55 http://docs.blueworx.com/BVR/InfoCenter/V6.1/help/index.jsp?topic=%2Fcom.ibm.wvraix. geninf.doc%2Fi658215.html
56 https://en.wikipedia.org/wiki/Loop_start
57 https://en.wikipedia.org/wiki/Ground_start
58 www.cisco.com/c/en/us/support/docs/voice/h323/14003-e-m-overview.html
59 https://tools.ietf.org/html/rfc5244#section-2.2
60 www.gaoresearch.com/PDF/MFR1R2.pdf
61 www.britishtelephones.com/pwover1.htm
62 https://en.wikipedia.org/wiki/Signaling_System_No._5
63 www.itu.int/rec/dologin_pub.asp?lang=s&id=T-REC-Q.7-198811-I!!PDF-E&type=items
64 www.sciencedirect.com/topics/computer-science/signaling-system
65 www.eolss.net/Sample-Chapters/C05/E6-108-04.pdf
66 https://en.wikipedia.org/wiki/Strowger_switch
67 https://patents.google.com/patent/US638249
68 www.ericsson.com/en/about-us/history/products/the-switches/the-crossbar-switch–from-concept-to-success
69 www.daenotes.com/electronics/communication-system/digital-switching
70 www.tutorialspoint.com/telecommunication_switching_systems_and_networks/telecommuni cation_switching_systems_and_networks_time_division_switching.htm
71 www.tutorialspoint.com/circuit-switching
72 www.tutorialspoint.com/packet-switching
73 www.techopedia.com/definition/20681/message-switching
74 http://denninginstitute.com/modules/atm/ATMswitch.html
75 https://ecomputernotes.com/computernetworkingnotes/communication-networks/what-is-data-communication
76 www.analog.com/en/analog-dialogue/articles/analog-to-digital-converter-architectures-and-choices.html#
77 https://en.wikipedia.org/wiki/Digital-to-analog_converter#Types
78 https://us.flukecal.com/blog/what-hart-protocol
79 https://en.wikipedia.org/wiki/Modem
80 https://circuitdigest.com/tutorial/serial-communication-protocols
81 www.omega.co.uk/techref/das/rs-232-422-485.html
82 www.electronics-notes.com/articles/connectivity/usb-universal-serial-bus/basics-tutorial.php
83 www.cypress.com/file/134171/download

84   www.beyondlogic.org/usbnutshell/usb3.shtml
85   www.arcelect.com/X21_interface.htm
86   www.fiberoptics4sale.com/blogs/archive-posts/95045126-what-is-pulse-code-modulation-pcm
87   www.yourdictionary.com/nyquist-theorem
88   www.techopedia.com/definition/9669/time-division-multiplexing-tdm
89   https://en.wikipedia.org/wiki/Digital_Private_Network_Signalling_System
90   http://samsiev.eu/Users/ATC/DPNSS/E20_DASS2_DPNSS.pdf
91   www.tlc-direct.co.uk/Technical/Telecoms/History/TH022.htm
92   www.idc-online.com/technical_references/pdfs/data_communications/What_is_a_QSIG.pdf
93   www.ciscopress.com/articles/article.asp?p=29737&seqNum=3
94   https://en.wikipedia.org/wiki/G.711
95   https://en.wikipedia.org/wiki/G.729
96   https://en.wikipedia.org/wiki/%CE%9C-law_algorithm
97   www.freesoft.org/CIE/Topics/126.htm
98   www.patton.com/whitepapers/intro_to_ss7_tutorial.pdf
99   www.mpirical.com/glossary/mtp-message-transfer-part
100 www.cs.rutgers.edu/~rmartin/teaching/fall04/cs552/readings/ss7.pdf
101 https://wiki.wireshark.org/MTP3
102 http://docs.blueworx.com/BVR/InfoCenter/V6.1/help/index.jsp?topic=%2Fcom.ibm.wvraix.
     newss7.doc%2Fthess7protocolstac3.html
103 https://en.wikipedia.org/wiki/ISDN_User_Part
104 www.telecomspace.com/ss7-sccp.html
105 https://telecomprotocols.blogspot.com/2012/09/ss7-protocol-stack-sccp.html
106 https://en.wikipedia.org/wiki/Transaction_Capabilities_Application_Part
107 www.techopedia.com/definition/30102/intelligent-network-application-part-inap
108 www.itu.int/rec/dologin_pub.asp?lang=e&id=T-REC-Q.1219-199404-I!!PDF-E&type=items
109 www.cellsoft.de/telecom/v5.htm
110 www.geeksforgeeks.org/isdn-protocol-architecture/
111 www.geeksforgeeks.org/digital-subscriber-line-dsl/
112 www.sciencedirect.com/topics/engineering/digital-subscriber-lines
113 www.sciencedirect.com/topics/computer-science/voice-over-internet-protocol
114 www.cse.wustl.edu/~jain/cis788-99/ftp/voip_protocols/
115 https://cdn.ttgtmedia.com/searchVoIP/downloads/Building_a_VoIP_Network_Ch[1]._8.pdf
116 www.3cx.com/pbx/sdp/
117 www.csie.ntu.edu.tw/~acpang/course/voip_2003/slides/Chap4_PartII.pdf
118 http://software.sonicwall.com/applications/app/index.asp?ev=appd&appid=1299&app_name=
     H.225%20Call%20Signaling
119 www.researchgate.net/figure/T120-protocol-stack-The-T120-protocol-stack-shown-in-
     Figure-3-is-based-on-a-layered_fig3_2618112
120 www.geeksforgeeks.org/real-time-transport-protocol-rtp/
121 www.geeksforgeeks.org/real-time-transport-control-protocol-rtcp/
122 https://en.wikipedia.org/wiki/Inter-Asterisk_eXchange
123 https://en.wikipedia.org/wiki/Skype_protocol
124 https://support.huawei.com/huaweiconnect/carrier/en/thread-76319-1-1.html
125 www.efort.com/media_pdf/GCP_EFORT_ENG.pdf
126 https://en.wikipedia.org/wiki/Media_Gateway_Control_Protocol
127 www.itu.int/ITU-D/treg/Events/Seminars/2010/Ghana10/pdf/Session2.pdf
128 https://en.wikipedia.org/wiki/Internet_Protocol
129 www.guru99.com/ip-header.html
130 www.paessler.com/it-explained/ip-address
131 http://www-sop.inria.fr/members/Vincenzo.Mancuso/ReteInternet/06_tcp_part1.pdf
132 www.tutorialspoint.com/data_communication_computer_network/transmission_control_
     protocol.htm
133 www.tutorialspoint.com/data_communication_computer_network/user_datagram_protocol.htm

134 www.geeksforgeeks.org/user-datagram-protocol-udp/
135 https://developer.mozilla.org/en-US/docs/Web/HTTP/Overview
136 https://developer.mozilla.org/en-US/docs/Web/HTTP/Methods
137 https://developer.mozilla.org/en-US/docs/Web/HTTP/Headers
138 www.manageengine.com/network-monitoring/what-is-snmp.html
139 www.ibm.com/support/knowledgecenter/SSGU8G_14.1.0/com.ibm.snmp.doc/ids_snmp_010.htm
140 www.researchgate.net/publication/273830243_SMTP_Simple_Mail_Transfer_Protocol
141 www.interserver.net/tips/kb/mime-multi-purpose-internet-mail-extensions/
142 www.geeksforgeeks.org/multipurpose-internet-mail-extension-mime-protocol/
143 www.geeksforgeeks.org/file-transfer-protocol-ftp-in-application-layer/
144 https://afteracademy.com/blog/what-is-ftp-and-how-does-an-ftp-work
145 www.csun.edu/~jeffw/Semesters/2006Fall/COMP429/Presentations/Ch25-FTP.pdf
146 http://people.na.infn.it/~garufi/didattica/CorsoAcq/Trasp/Lezione9/tcpip_ill/tftp_tri.htm
147 www.ietf.org/rfc/rfc1939.txt
148 www.2brightsparks.com/resources/articles/understanding-post-office-protocol-pop3.pdf
149 www.mayan.cn/IA/15/8-TELNET-20150414.pdf
150 https://tools.ietf.org/html/rfc854
151 https://en.wikipedia.org/wiki/XMPP
152 https://tools.ietf.org/id/draft-ietf-xmpp-3921bis-01.html
153 https://en.wikipedia.org/wiki/Jingle_(protocol)
154 https://xmpp.org/extensions/xep-0166.html
155 www.guru99.com/routing-protocol-types.html
156 www.ciscopress.com/articles/article.asp?p=2180210&seqNum=7
157 www.geeksforgeeks.org/difference-between-classful-routing-and-classless-routing/
158 www.javatpoint.com/distance-vector-routing-algorithm
159 www.geeksforgeeks.org/unicast-routing-link-state-routing/
160 https://en.wikipedia.org/wiki/Path-vector_routing_protocol
161 https://en.wikipedia.org/wiki/Wireless
162 https://en.wikipedia.org/wiki/Wi-Fi
163 https://en.wikipedia.org/wiki/IEEE_802.11
164 https://core.ac.uk/download/pdf/234677258.pdf
165 http://ece-research.unm.edu/controls/papers/Jor_CTA.pdf
166 www.oreilly.com/library/view/getting-started-with/9781491900550/ch01.html
167 https://en.wikipedia.org/wiki/Zigbee
168 www.eetimes.com/6lowpan-the-wireless-embedded-internet-part-3-6lowpan-architecture-protocol-stack-link-layers/#
169 https://cdn.rohde-schwarz.com/pws/dl_downloads/dl_application/application_notes/1ma182/1MA182_5E_NFC_WHITE_PAPER.pdf
170 https://en.wikipedia.org/wiki/Customized_Applications_for_Mobile_networks_Enhanced_Logic
171 www.gl.com/camel-application-part-cap-emulator-over-tdm-ip-using-maps.html
172 www.tutorialspoint.com/wavelength-division-multiplexing
173 www.techopedia.com/definition/3451/wavelength-division-multiplexing-wdm
174 https://arxiv.org/pdf/1506.04836.pdf
175 www.itu.int/rec/dologin_pub.asp?lang=f&id=T-REC-G.651.1-201811-I!!PDF-E&type=items
176 www.fiber-optic-solutions.com/single-mode-fiber-difference.html
177 www.itu.int/rec/dologin_pub.asp?lang=e&id=T-REC-G.983.1-200501-I!!PDF-E&type=items
178 https://en.wikipedia.org/wiki/G.984
179 www.techopedia.com/definition/21314/plesiochronous-digital-hierarchy-pdh
180 www.ee.columbia.edu/~bbathula/courses/HPCN/chap04_part-2.pdf
181 https://en.wikipedia.org/wiki/Synchronous_optical_networking
182 www.metaswitch.com/knowledge-center/reference/what-is-optical-transport-network-otn
183 www.tutorialspoint.com/fiber-distributed-data-interface-fddi

184   www.w3schools.in/cyber-security/network-protocols-and-its-security/
185   www.hostinger.com/tutorials/ssh-tutorial-how-does-ssh-work
186   https://en.wikipedia.org/wiki/HTTPS
187   www.geeksforgeeks.org/secure-socket-layer-ssl/
188   www.geeksforgeeks.org/ip-security-ipsec/
189   www.simplilearn.com/what-is-kerberos-article
190   https://tools.ietf.org/html/rfc3711
191   www.speedcheck.org/wiki/wep/
192   www.geeksforgeeks.org/wifi-protected-access-wpa/
193   https://networklessons.com/cisco/ccna-200-301/wi-fi-protected-access-wpa
194   https://en.wikipedia.org/wiki/Extensible_Authentication_Protocol
195   https://tools.ietf.org/html/rfc5216
196   https://docs.microsoft.com/en-us/openspecs/windows_protocols/ms-peap/a128a089-0919-
       41a5-a0c2-9f25ef28289d
197   www.internetsociety.org/deploy360/tls/basics/
198   https://tools.ietf.org/html/rfc1827
199   www.techopedia.com/definition/26196/layer-2-tunneling-protocol-l2tp
200   www.sciencedirect.com/topics/computer-science/fabrication-attack
201   www.diva-portal.org/smash/record.jsf?pid=diva2%3A532677&dswid=-9724
202   www.tutorialspoint.com/radius/what_is_radius.htm
203   https://docs.microsoft.com/en-us/windows/win32/nps/internet-authentication-service-vs-
       network-policy-server
204   www.cisco.com/c/en/us/products/security/vpn-endpoint-security-clients/what-is-vpn.html#~
       how-a-vpn-works
205   www.netmotionsoftware.com/blog/connectivity/vpn-protocols
206   https://en.wikipedia.org/wiki/Remote_Desktop_Services
207   www.networkworld.com/article/3268449/what-is-dns-and-how-does-it-work.html
208   www.guru99.com/address-resolution-protocol.html
209   https://en.wikipedia.org/wiki/Address_Resolution_Protocol
210   www.ietf.org/rfc/rfc1305.txt
211   https://en.wikipedia.org/wiki/Network_Time_Protocol
212   https://en.wikipedia.org/wiki/Digital_privacy
213   https://seedscientific.com/how-much-data-is-created-every-day/
214   www.lifewithalacrity.com/2004/04/four_kinds_of_p.html
215   www.wired.com/story/zoombomb-inside-jobs/#:~:text=Even%20after%20Zoom%20
       password%2Dprotected,%2Dcalled%20Zoom%2Dbombing%20continued.&text=The%20
       phenomenon%20is%20explained%20in,high%20school%20and%20college%20classes.
216   https://us.norton.com/internetsecurity-wifi-what-is-a-man-in-the-middle-attack.html
217   https://com563.ua.edu/index.php/Communication_Privacy_Management_Theory
218   https://en.wikipedia.org/wiki/Privacy
219   https://scholarship.law.upenn.edu/cgi/viewcontent.cgi?article=1938&context=jil
220   www.rogerclarke.com/DV/Privacy.html
221   https://selfkey.org/data-breaches-in-2019/

Index

Printed in the United States
by Baker & Taylor Publisher Services